LOCALLY COMPACT SEMI-ALGEBRAS
with applications to spectral theory of positive operators

NORTH-HOLLAND
MATHEMATICS STUDIES 9

Locally Compact Semi-Algebras

with applications to spectral theory of positive operators

M. A. KAASHOEK
Vrije Universiteit, Amsterdam

and

T. T. WEST
Trinity College, Dublin

1974

NORTH-HOLLAND PUBLISHING COMPANY - AMSTERDAM • LONDON
AMERICAN ELSEVIER PUBLISHING COMPANY, INC. - NEW YORK

Library of Congress Catalog Number: 73–92553
ISBN North-Holland:
Series: 0 7204 2700 2
Volume: 0 7204 2709 6
ISBN American Elsevier: 0 444 10609 x

PUBLISHERS:

NORTH-HOLLAND PUBLISHING COMPANY – AMSTERDAM
NORTH-HOLLAND PUBLISHING COMPANY, LTD.–LONDON

SOLE DISTRIBUTORS FOR THE U.S.A. AND CANADA:

AMERICAN ELSEVIER PUBLISHING COMPANY, INC.
52 VANDERBILT AVENUE
NEW YORK, N.Y. 10017

PRINTED IN THE NETHERLANDS

PREFACE

In this monograph we examine the connection between the structure of locally compact semi-algebras and spectral theory, in particular, the spectral theory of positive linear operators.

A semi-algebra is a subset of a Banach algebra which is closed under addition, multiplication and multiplication by non-negative real numbers. Local compactness of Banach algebras implies finite dimensionality and it follows that the spectrum of any element therein consists of a finite set of poles. This from our point of view is the trivial case. However, locally compact semi-algebras may contain elements whose spectrum is not trivial and this fact gives them an important role in the spectral theory of compact positive operators.

F.F. Bonsall first employed locally compact semi-algebras to algebraicise the Frobenius theory of matrices with non-negative entries (see section 7 of [2]). Bonsall and Tomiuk [4] used locally compact semi-algebras to put the spectral theory of compact positive operators in an algebraic setting and to extend the Krein-Rutman theory of compact irreducible positive operators. This was made possible by Bonsall's work (analogous to the Wedderburn theory) showing the existence of idempotents and minimal ideals in locally compact semi-algebras (see section 3 of [2]). The existence of idempotents is crucial in spectral theory and provides the link between algebra and analysis.

In [4] Bonsall and Tomiuk proved that a compact linear operator of unit spectral radius whose spectrum contains the point one generates a locally compact semi-algebra. Our work started with an attempt to find a converse to this theorem. This was obtained in the case of an operator with equibounded iterates ([19], Theorem 5). It emerged that in this case the local compactness of the semi-algebra depends on the fact that the closed semigroup generated by the operator is compact, and further that this semigroup is intimately connected with the spectral properties of its generator. Continued investigation led to the appearance of many links between semigroups, semi-algebras and spectral theory. This monograph represents an attempt to collect these results and set them out in a logical and readable manner.

Detailed information on the contents of the monograph is given in the introductory sections of the various chapters. The chief prerequisite for reading this monograph is a knowledge of spectral theory of elements of Banach algebras and of linear operators. The spectral theory which we use is outlined in a preliminary chapter. With the exception of sections 1, 3, 4 and 5 of Chapter I the only topology we use is the norm topology. Sections 4 and 5 of Chapter I make use of the weak operator topology on the bounded linear operators on a Hilbert space and of the spectral theory of unitary operators. These sections contain the construction of an important counter-example in semigroup theory but are not crucial to the remainder of the text.

Each chapter concludes with notes which cover related material which is not dealt with in the text, give further references and pose some unsolved problems.

The lemmas, propositions, theorems and corollaries are numbered serially in each section of each chapter. Thus Theorem II.1.2 is the second result of section 1 of Chapter II. Within Chapter II this result is referred to as Theorem 1.2. Proposition P.6 refers to Proposition 6 in the preliminary chapter. A similar system is used for the numbering of equations and formulae.

The authors express their gratitude to Professor A.C. Zaanen for his advice and criticism and to Miss Betty Kuiper who typed the monograph.

M.A. Kaashoek

T.T. West

CONTENTS

LIST OF SYMBOLS AND NOTATIONS

$\emptyset$	: empty set
$\mathbb{C}$	: complex plane
$\mathbb{D}$	: closed unit disk in $\mathbb{C}$
Γ	: unit circle in $\mathbb{C}$
$\mathbb{R}$	: real numbers
$\mathbb{R}^+(\mathbb{R}_o^+)$	: strictly positive (non-negative) real numbers
$\mathbb{Z}$	: integers
$\mathbb{Z}^+(\mathbb{Z}^-)$	: strictly positive (negative) integers
clQ	: closure of a subset Q of a topological space
coQ	: convex hull of a subset Q of a linear space
K	: cone in a linear space
H	: Hilbert space
E	: Banach space
T	: bounded linear operator E
L(E)	: Banach algebra of bounded linear operators on E
B	: complex Banach algebra with unit element e
A	: semi-algebra in B
t	: element of B
t^{-1}	: inverse of t
$\sigma(t)$	: spectrum of t
$\mathrm{Per}\sigma(t)$	: peripheral spectrum of t
r(t)	: spectral radius of t
$\rho(t)$	: resolvent set of t
R(z;t)	: resolvent element $(ze - t)^{-1}$
$\Lambda(t)$	: semigroup $\{t^n: n \in \mathbb{Z}^+\}$
S(t)	: closure of $\Lambda(t)$
K	: kernel of S(t)
A(t)	: closed semi-algebra generated by t

PRELIMINARIES

This preliminary chapter consists of a collection of definitions and results from spectral theory in Banach algebras which are used throughout the text. Most of the results are given without proofs but with adequate references.

Let B be a complex Banach algebra with unit e. We assume that $\|e\| = 1$. An element $t \in B$ is *invertible* if there exists an element $t^{-1} \in B$ satisfying $t^{-1}t = tt^{-1} = e$. The element t^{-1} is the *inverse* of t and is unique.

Let $t \in B$. The *resolvent set* $\rho(t)$ of t is the set of all complex numbers λ such that $\lambda e - t$ is invertible in B. For $\lambda \in \rho(t)$ the element $(\lambda e - t)^{-1}$ is called the *resolvent* of t at λ. Often it will be denoted by $R(\lambda;t)$. The resolvent set is an open subset of the complex plane $\mathbb{C}$, and on this set the function

$$\lambda \to R(\lambda;t)$$

is locally analytic ([16], Theorem 4.7.1). In fact, for $\lambda_o \in \rho(t)$ and $|\lambda - \lambda_o| < \|R(\lambda_o;t)\|^{-1}$, we have

$$R(\lambda;t) = \sum_{n=o}^{\infty} (\lambda_o - \lambda)^n R(\lambda_o;t)^{n+1}. \tag{1}$$

The series (1) converges in the norm of B.

The complement of $\rho(t)$ in $\mathbb{C}$ is called the *spectrum* of t and is denoted by $\sigma(t)$. It is a non-empty compact subset of $\mathbb{C}$ ([16], Theorem 4.7.4). The *spectral radius* of t, denoted by $r(t)$, is defined by

$$r(t) = \max \{|\lambda| : \lambda \in \sigma(t)\}.$$

It has the following properties ([16], Theorem 4.7.3).

1. PROPOSITION. *For each* t *in* B, *we have*

(i) $r(t) = \lim_{n\to\infty} \|t^n\|^{1/n}$;

(ii) $r(t) \leq \|t\|$;

(iii) $r(t^k) = r(t)^k \quad (k = 1,2,\ldots)$;

(iv) $r(\alpha t) = |\alpha| r(t) \quad (\alpha \in \mathbb{C})$.

We shall use the term *peripheral spectrum* of t to denote the set

$$\mathrm{Per}\sigma(t) = \{\lambda \in \sigma(t) : |\lambda| = r(t)\}.$$

For $|\lambda| > r(t)$, we have $\lambda \in \rho(t)$ and

$$R(\lambda;t) = \sum_{n=0}^{\infty} \lambda^{-(n+1)} t^n.$$

This series, convergent in norm, is called the *Neumann series* for t ([16], Theorem 4.7.2).

2. PROPOSITION. *Let* V *be a commutative subset of* B, *and let* $s,t \in V$ *Then*

(i) $r(s + t) \leq r(s) + r(t)$;

(ii) $r(st) \leq r(s)r(t)$;

(iii) r *is a continuous function on* V.

PROOF. For (i) and (ii) see [28], Theorem 1.4.1(v). To prove (iii), observe that (i) implies that

$$|r(t) - r(s)| \leq r(t - s).$$

By combining this with Proposition 1(ii), we see that

$$|r(t) - r(s)| \leq \|t - s\|,$$

and hence r is continuous on V.

An element $t \in B$ is *nilpotent of order* k if $t^k = 0$ and if k is the smallest positive integer for which this is true. An element $p \in B$ is an *idempotent* if $p^2 = p$. It is easy to see that for a non-zero idempotent p the spectral radius $r(p) = 1$.

Let $t \in B$, and let f be a complex valued function defined and locally analytic on an open neighbourhood of $\sigma(t)$. The element $f(t)$ is defined by

$$f(t) = \frac{1}{2\pi i} \int_\Gamma f(\lambda)R(\lambda;t)d\lambda,$$

where Γ is a suitable contour surrounding $\sigma(t)$ (see [9], Definition VII.3.9 for details). The mapping

$$f \to f(t) \qquad (2)$$

of functions locally analytic in a neighbourhood of $\sigma(t)$ into B is an algebraic homomorphism which preserves polynomials.

The next result is known as the spectral mapping theorem.

3. PROPOSITION. *Let* f *be a complex valued function which is locally analytic on a neighbourhood of* $\sigma(t)$. *Then*

$$\sigma(f(t)) = f(\sigma(t)) = \{f(\lambda) : \lambda \in \sigma(t)\}.$$

PROOF. See [9], Theorem VII.3.11.

Let $t \in B$. A subset σ of $\sigma(t)$ which is both open and closed in the relative topology of $\sigma(t)$ is called a *spectral set* of t. Let σ be a spectral set of t, and consider the function h which is defined to be one on an open neighbourhood of σ and to be zero on an open neighbourhood of $\sigma(t) \setminus \sigma$. Then h is locally analytic on a neighbourhood V of $\sigma(t)$ and

$$h(\lambda)^2 = h(\lambda) \quad (\lambda \in V).$$

Since the mapping (2) is a homomorphism, it follows that $h(t)^2 = h(t)$, i.e., $h(t)$ is an idempotent in B. We call $h(t)$ the *spectral idempotent* of t corresponding to the spectral set σ. Clearly, $h(t) = e$ if $\sigma = \sigma(t)$ and $h(t) = 0$ if $\sigma = \emptyset$.

4. PROPOSITION. *Let* f *be a complex valued function which is locally analytic on a neighbourhood of* $\sigma(t)$. *Let* σ *be a spectral set of* t, *and let* p *be the corresponding spectral idempotent. Then*

(i) $\sigma(f(t)p) = \{f(\lambda) : \lambda \in \sigma\} \cup \{0\}$ *if* $\sigma \neq \sigma(t)$;

(ii) $r(f(t)p) = \sup\{|f(\lambda)| : \lambda \in \sigma\}$ *if* $\sigma \neq \emptyset$.

PROOF. Use the definition of p together with the fact that the mapping (2) is a homomorphism and apply the spectral mapping theorem.

5. PROPOSITION. *If* σ *is a spectral set of* t *and* $\sigma \neq \sigma(t)$, *then* $p \neq e$ *and*

(i) $\sigma(tp) = \sigma \cup \{0\}$,

(ii) $R(\lambda;tp) = R(\lambda;t)p + \frac{1}{\lambda}(e - p)$ *for* $0 \neq \lambda \in \rho(t)$.

PROOF. (i) is a special case of Proposition 4(i), and (ii) follows by an elementary computation.

If p is a spectral idempotent for t, then $p = f(t)$ for a suitable function f. Conversely, if q is an idempotent in B and if $q = f(t)$ for some locally analytic function f, then q is a spectral idempotent. In fact the following more general theorem holds (see [8], Theorem 3.1 or [9],

Exercise VII.8.3). As the result is used regularly in the monograph, we include a proof.

6. PROPOSITION. *Let* $\{f_n\}$ *be a sequence of complex valued functions which are locally analytic on some (not necessarily fixed) neighbourhood of* $\sigma(t)$. *Let* q *be an idempotent in* B *such that*

$$\lim_n f_n(t) = q$$

in the norm of B. *Then* q *is a spectral idempotent.*

PROOF. Let $t \in B$. Using Propositions 1(ii) and 3 together with the fact that the mapping (2) is a homomorphism, one sees that

$$\begin{aligned} |f_n(\lambda) - f_m(\lambda)| &\leq r(f_n(t) - f_m(t)) \\ &\leq \|f_n(t) - f_m(t)\|. \end{aligned}$$

Thus $\{f_n\}$ is a Cauchy sequence of continuous functions on the compact set $\sigma(t)$ in the supremum norm. It follows that there exists a continuous function g defined on $\sigma(t)$ such that

$$f_n(\lambda) \to g(\lambda) \quad (n \to +\infty) \tag{3}$$

uniformly on $\sigma(t)$.

The hypothesis that q is an idempotent implies that

$$f_n(t)^2 - f_n(t) \to 0 \quad (n \to +\infty).$$

Therefore, as above,

$$f_n(\lambda)^2 - f_n(\lambda) \to 0 \quad (n \to +\infty)$$

uniformly on $\sigma(t)$. But then (3) implies that

$$g(\lambda)^2 = g(\lambda) \quad (\lambda \in \sigma(t)). \tag{4}$$

Write

$$\sigma = \{\lambda \in \sigma(t) : g(\lambda) = 1\}.$$

Then, by (4) and the continuity of g, σ is an open and closed subset of $\sigma(t)$ and is therefore a spectral set of t.

Let p be the corresponding spectral idempotent. We shall show that $q = p$. Suppose that $\sigma \neq \emptyset, \sigma(t)$. By Proposition 4(ii)

$$r(f_n(t)(e - p)) = \sup\{|f_n(\lambda)| : \lambda \in \sigma(t) \setminus \sigma\}$$

and

$$r((f_n(t) - e)p) = \sup \{|f_n(\lambda) - 1| : \lambda \in \sigma\}.$$

By (3), (4) and the definition of σ, the sequence $\{f_n(\lambda)\}$ converges uniformly to zero on $\sigma(t) \setminus \sigma$ and to one on σ. Thus

$$\lim_n r(f_n(t)(e - p)) = 0, \quad \lim_n r((f_n(t) - e)p) = 0.$$

Therefore by the continuity of r on commutative subsets of B, we have

$$r(q(e - p)) = 0, \quad r((q - e)p) = 0.$$

But $q(e - p)$ and $(q - e)p$ are idempotents, therefore both are zero. Hence $q = qp = p$ and q is a spectral idempotent.

A similar argument shows that if $\sigma = \sigma(t)$, then $q = e$ and if $\sigma = \emptyset$, then $q = 0$. In both cases q is a spectral idempotent.

Let $t \in B$. A point $\lambda_0 \in \sigma(t)$ is an *isolated point* in $\sigma(t)$ if $\{\lambda_0\}$ is a spectral set of t. Suppose that λ_0 is such a point. Then the function

$$\lambda \to R(\lambda;t) \tag{5}$$

is defined and analytic in a deleted neighbourhood of λ_0. Thus there exists a Laurent expansion of $R(\lambda;t)$ in powers of $\lambda - \lambda_0$ and with coefficients in B, say

$$R(\lambda;t) = \sum_{n=-\infty}^{\infty} (\lambda - \lambda_0)^n a_n. \tag{6}$$

We have the following information (see [32], §5.8).

7. PROPOSITION. (i) a_{-1} *is the spectral idempotent corresponding to the spectral set* $\{\lambda_0\}$;

(ii) $(t - \lambda_0 e)a_0 = a_{-1} - e$;

(iii) $(t - \lambda_0 e)a_{n+1} = a_n$ *for* $n \neq -1$.

Let λ_0 be an isolated point of $\sigma(t)$, and let the Laurent expansion of $R(\lambda;t)$ in a deleted neighbourhood of λ_0 be given by (6). The point λ_0 is a *pole* of t of *order* k (≥ 1) if

$$a_{-k} \neq 0, \quad a_n = 0 \quad (n < -k).$$

In other words, λ_0 is a pole of t of order k if λ_0 is a pole of order k of the vector valued function (5). If $k = 1$, then λ_0 is called a *simple pole* of t.

8. PROPOSITION. *Suppose that* $\sigma(t)$ *consists of poles of* t. *Then the subalgebra of* B *generated by* t *and* e *is finite dimensional.*

PROOF. Since poles are isolated points and $\sigma(t)$ is compact, $\sigma(t)$ is a finite set,

$$\sigma(t) = \{\lambda_1, \lambda_2, \ldots, \lambda_r\}$$

say. Let k_i be the order of the pole λ_i, and let p_i be the spectral idempotent corresponding to the spectral set $\{\lambda_i\}$. Then $p_1 + p_2 + \ldots + p_r = e$, and, by Proposition 7(i) and (ii),

$$(t - \lambda_i e)^{k_i} p_i = 0 \quad (i = 1,2,\ldots,r).$$

Since $(t - \lambda_i e)^{k_i}$ and p_j commute for all i and j, this implies that

$$\prod_{i=1}^{r} (t - \lambda_i e)^{k_i} = \sum_{j=1}^{r} \left(\prod_{i=1}^{r} (t - \lambda_i e)^{k_i} p_j\right) = 0.$$

Thus t is a zero of a non-zero polynomial, and hence the result follows.

Let $t \in B$. We say that t has *equibounded iterates* whenever there exists $M \geq 0$ such that $\|t^n\| \leq M$ for $n = 1,2,\ldots$. If t is such an element, then $r(t) \leq 1$.

9. PROPOSITION. *Suppose that* t *has equibounded iterates, and let* λ *be a pole of* t. *If* $|\lambda| = 1$, *then* λ *is a simple pole of* t.

PROOF. Let $\lambda = e^{i\phi}$, and choose $z = re^{i\phi}$ with $r > 1$. Then the Neumann series gives

$$\begin{aligned}\|(z - \lambda)R(z;t)\| &= (r - 1)\left\|\sum_{n=0}^{\infty} z^{-(n+1)} t^n\right\| \\ &\leq (r - 1) \sum_{n=0}^{\infty} M\, r^{-(n+1)} \\ &= M.\end{aligned}$$

Therefore $(z - \lambda)R(z;t)$ remains bounded for $z = re^{i\phi}$ and $r > 1$. Now the absolute value $|f(z)|$ of a complex valued locally analytic function f tends to $+\infty$ if z approaches a pole of f. Hence, by using the Hahn-Banach theorem, it follows that the function

$$z \to (z - \lambda)R(z;t)$$

does not have a pole at λ, and thus λ is a simple pole of t.

Let E be a complex Banach space. Denote by L(E) the complex Banach algebra of all bounded linear operators on E with unit I the identity operator on E. An idempotent in L(E) is called a *projection* of E.

Let $T \in L(E)$. The spectral properties of T are those obtained by considering T as an element of the Banach algebra L(E). Since E is a Banach space, $\lambda \in \rho(T)$ if and only if the map $\lambda I - T$ is bijective. A point $\lambda \in \mathbb{C}$ is an *eigenvalue* of T if there exists $x \neq 0$ in E such that $Tx = \lambda x$. The eigenvalues of T form a subset of $\sigma(T)$ called the *point spectrum* of T. The *continuous spectrum* of T consists of those λ in $\sigma(T)$ such that λ is not an eigenvalue of T and $(\lambda I - T)E$ is a proper dense subspace of E.

Let λ be a pole of T of order k, and let P be the corresponding spectral projection. Proposition 7 shows that

$$(T - \lambda I)^{k-1}P \neq 0, \quad (T - \lambda I)^{k}P = 0.$$

Choose $u \in E$ such that $x = (T - \lambda I)^{k-1}Pu \neq 0$. Then

$$Tx - \lambda x = (T - \lambda I)x = 0.$$

Hence λ is an eigenvalue of T.

Let λ be a pole of T, and let P be the corresponding spectral projection. We call λ a pole of *finite rank* if the range of P is finite dimensional.

An element T in L(E) is *compact* if T maps the unit ball of E into a compact subset of E.

10. PROPOSITION. *If* $T \in L(E)$ *is compact, then*

$$\sigma(T) \setminus \{0\}$$

consists of poles of T *of finite rank.*

PROOF. See [9], Theorem VII.4.5.

CHAPTER I

COMPACT MONOTHETIC SEMIGROUPS

In Chapter I we examine the structure of compact monothetic semigroups and its connection with spectral theory. Starting with the case of jointly continuous multiplication in section 1, the structure of compact monothetic semigroups is completely described in Theorem 1.1. There we obtain the existence of a unique minimal ideal which is a compact monothetic group whose unit is the only idempotent in the semigroup (Numakura [24]). Methods from ergodic theory are used in section 2 to obtain Theorem 2.3 which characterizes compact monothetic semigroups.in Banach algebras in terms of the spectrum of a generating element. It states that the closed monothetic semigroup generated by an element is compact if, and only if, the spectrum of the generator is contained in the closed unit disk and its intersection with the unit circle is a finite (possibly empty) set of simple poles.

The structure of compact monothetic semigroups is much more complex if multiplication is only separately continuous. An important example of such a semigroup in infinite dimensional Hilbert space is the weak operator closed monothetic semigroup generated by a linear operator with equibounded iterates. The structure theory of such semigroups is due to de Leeuw and Glicksberg [23]. In section 3 we obtain the existence of a unique minimal ideal which is a compact topological group. The representation theory of weakly compact groups gives information on the spectral properties of the generator (Theorems 4.1 and 4.2). However, one does not have a unique idempotent in such a semigroup (as in the case of jointly continuous multiplication). In Theorem 5.1 we construct a unitary operator in Hilbert space which generates a weak operator compact semigroup with at least two idempotents. More detailed analysis in Theorem 5.2 shows that, in fact, this semigroup contains uncountably many distinct idempotents. These results are due to West [33] and Brown and Moran [5]. Further pathology of the idempotent sets of such semigroups has been exhibited by Brown and Moran in [6] and [7].

1. JOINTLY CONTINUOUS MULTIPLICATION

Let Ω be a topological semigroup with *jointly continuous multiplication*, i.e., Ω is a semigroup endowed with a Hausdorff topology such that the map

$$(x,y) \to xy$$

from $\Omega \times \Omega$ into Ω is continuous. If $t \in \Omega$, denote by $\Lambda(t)$ the set

$$\{t^n : n = 1,2,\ldots\},$$

and by $S(t)$ the closure of $\Lambda(t)$ in Ω. Joint continuity of multiplication implies at once that $S(t)$ is a commutative semigroup. The semigroup Ω is said to be *monothetic* whenever there exists $t \in \Omega$ such that $\Omega = S(t)$.

Let $S(t)$ be a compact monothetic semigroup with jointly continuous multiplication. The structure theory of $S(t)$ is originally due to Numakura [24]; a good account is given by Hewitt and Ross ([15], §8). A non-empty subset I of $S(t)$ is an *ideal* of $S(t)$ if

$$IS(t) \subset I.$$

The key to the structure of $S(t)$ is the existence of a minimal ideal K (with respect to inclusion) called the *kernel* of $S(t)$ which turns out to be a compact topological group.

1.1 THEOREM. *Let* $S(t)$ *be a compact monothetic semigroup with jointly continuous multiplication. Then*

(i) $K = \bigcap\{xS(t) : x \in S(t)\}$ *is the unique minimal ideal in* $S(t)$*;*

(ii) K *is a compact topological group;*

(iii) *the unit* e_1 *of* K *is the sole idempotent in* $S(t)$ *and*

$$K = e_1S(t) = S(e_1t).$$

PROOF. (i) The collection of ideals $\{xS(t) : x \in S(t)\}$ has the finite intersection property because

$$x_1x_2 \dots x_n \in \bigcap_{i=1}^{n} x_i\, S(t).$$

Continuity of multiplication implies that xS(t) is a closed subset of S(t) for each x. Since S(t) is compact both facts together imply that K is a non-empty closed subset of S(t). Clearly K is an ideal of S(t).

Let I be an ideal of S(t). Choose x in I, then

$$K \subset xS(t) \subset I.$$

Thus K is the unique minimal ideal of S(t).

(ii) We know already that K is a closed subset of S(t). Hence K is compact, because S(t) is compact.

Next we show that K is a group. Let $y \in K$. Then

$$yK = \bigcap\{yxS(t) : x \in S(t)\} \subset K.$$

But yK is an ideal, because $(yK)S(t) \subset yK$. Therefore, since K is minimal, $K \subset yK$. Hence

$$yK = K \quad (y \in K).$$

This implies that K is a semigroup. Further it shows that, for any y and z in K, the equation yX = z has a solution in K. Since multiplication in K is commutative, these facts together imply that K is a group.

To see that K is a topological group, we need to show that the map

$$x \to x^{-1}$$

from K into K is continuous. Let $\{x_\alpha\}$ be a net in K converging to x, then $\{x_\alpha^{-1}\}$ is also a net in K. Since K is compact it has a converging subnet. Let y in K be the limit of some convergent subnet of $\{x_{\alpha_1}^{-1}\}$. Joint continuity of multiplication shows that $xy = e_1$ and thus $y = x^{-1}$. Hence every convergent subnet converges to x^{-1} and so

$$x_\alpha^{-1} \to x^{-1}.$$

This concludes the proof of (ii).

(iii) For each n we have $e_1t^n = (e_1t)^n \in K$. This implies that

$$e_1S(t) \subset S(e_1t) \subset K.$$

But $e_1S(t)$ is an ideal of S(t). Therefore $K = e_1S(t)$, and hence $K = e_1S(t) = S(e_1t)$.

Let f be an idempotent in S(t), i.e., $f^2 = f$. We want to show that

$f = e_1$. Since $K = e_1S(t)$ is a group with unit e_1, it suffices to show that $f \in e_1S(t)$. Suppose not. Then, as $e_1S(t) = K$ is compact, there exist open sets U and V such that

$$f \in V, \quad e_1S(t) \subset U, \quad V \cap U = \emptyset.$$

Using the joint continuity of the multiplication in S(t), we see that the inclusion $e_1S(t) \subset U$ implies that for each x in S(t) there exist open sets O_x and U_x such that

$$e_1 \in O_x, \quad x \in U_x, \quad yz \in U \quad (y \in O_x,\ z \in U_x).$$

From the compactness of S(t) it follows that there exist $x_1,\ldots,x_n$ in S(t) such that

$$S(t) \subset U_{x_1} \cup U_{x_2} \cup \ldots \cup U_{x_n}.$$

Put $O = \bigcap_{i=1}^{n} O_{x_i}$. Then O is an open neighbourhood of e_1 and

$$OS(t) \subset U.$$

Since $e_1 \in S(t)$, there exists a positive integer m such that $t^m \in O$. Next we observe that f is a cluster point of the sequence $\{t^n\}$, i.e., there exists a subnet $\{n_\alpha\}$ of the sequence of positive integers such that

$$t^{n_\alpha} \to f \tag{1}$$

This follows from the fact that f is an idempotent in S(t). Indeed, if f does not belong to $\{t,t^2,\ldots\}$, then (1) is trivially true because, by assumption, $f \in S(t) = \text{cl}\ \{t,t^2,\ldots\}$. If $f = t^r$ for some positive integer r, then, using $f^2 = f$, we have

$$f = t^{2nr} \quad (n = 1,2,\ldots).$$

So in that case (1) also holds. Since $f \in V$, formula (1) implies that there exists $k > m$ such that $t^k \in V$. But then

$$t^k = t^m t^{k-m} \in OS(t) \subset U.$$

This contradicts the fact that $V \cap U = \emptyset$. Hence f belongs to $e_1S(t)$ and e_1 is the sole idempotent in S(t). This completes the proof.

Note that the arguments of the last part of the preceding proof can be used to show that each cluster point of the sequence $\{t^n\}$ is in K. Conversely any element in K is a cluster point of this sequence because e_1 is one and e_1 is the unit of K. So K is precisely the set of cluster points of the sequence $\{t^n\}$.

2. SEMIGROUPS IN BANACH ALGEBRAS

Let B be a complex Banach algebra with unit e. We suppose B to be endowed with the norm topology. Then multiplication in B is jointly continuous. So, if $t \in B$ and $S(t)$ is compact, the situation discussed in section 1 applies. We use some techniques known in ergodic theory (see Dunford [8]) to determine the spectral properties of the generating element t.

Recall (see Preliminaries) that we use the symbol $\sigma(t)$ to denote the spectrum of t and $r(t)$ to denote its spectral radius. Further a point λ in $\sigma(t)$ is called a simple pole of t if it is a pole of the resolvent function

$$z \to (ze - t)^{-1}.$$

2.1 PROPOSITION. *Let* $S(t)$ *be compact in* B. *Then*

$$t_n = \frac{1}{n}(t + t^2 + \dots + t^n) \to p \quad (n \to +\infty)$$

in the norm of B. *Here* p *is an idempotent in* B *such that*

$$(e - t)p = p(e - t) = 0.$$

Further $p \neq 0$ *if and only if* $1 \in \sigma(t)$, *and in that case* 1 *is a simple pole of* t.

PROOF. For each n in $\mathbf{Z}^+$, the set of positive integers, t_n belongs to the closed convex hull Σ of $S(t)$. By Mazur's theorem on the closed convex hull of a compact set, Σ is compact in B ([9], Theorem V.2.6). Therefore there exists p in B and an increasing sequence $\{n_i\}$ in $\mathbf{Z}^+$ such that

$$t_{n_i} \to p \quad (i \to +\infty).$$

Obviously

$$p(e - t) = (e - t)p.$$

Since Σ is compact, there exists a positive number M such that

$$\|t^n\| \leq M \quad (n \in \mathbf{Z}^+).$$

Hence, for each n in $\mathbf{Z}^+$,

$$\|t_n(e - t)\| = \|\tfrac{1}{n}(t - t^{n+1})\| \leq 2Mn^{-1}.$$

Thus $t^n(e - t) \to 0$ if $n \to +\infty$, and therefore

$$p(e - t) = 0. \tag{1}$$

The next step is to show that

$$t_n \to p \quad (n \to +\infty). \tag{2}$$

It follows from (1) that $t^n p = p$ for each n in $\mathbb{Z}^+$. Hence

$$ap = p \quad (a \in \Sigma). \tag{3}$$

Let q be another cluster point of the sequence $\{t^n\}$. From (3) we get

$$qp = p. \tag{4}$$

Interchanging the roles of p and q gives

$$pq = q.$$

But Σ is commutative, therefore $p = q$ and the compactness of Σ implies (2). Putting $q = p$ in (4) shows that p is an idempotent.

Suppose $1 \in \sigma(t)$. Then, by the spectral mapping theorem (Proposition P.3), $1 \in \sigma(t_n)$ for each n in $\mathbb{Z}^+$. Thus

$$\|t_n\| \geq r(t_n) \geq 1 \quad (n \in \mathbb{Z}^+).$$

Continuity gives $\|p\| \geq 1$. In particular, $p \neq 0$.

Conversely, if $p \neq 0$, then (1) implies $e - t$ is not invertible in B, i.e., $1 \in \sigma(t)$.

Since p is the limit in B of polynomials in t, we know that p is a spectral idempotent associated with a spectral set of t (see Proposition P. 6). If $p \neq 0$, equation (1) shows that this spectral set consists solely of the point 1. Hence 1 is an isolated point of $\sigma(t)$. Using (1) again, it follows that 1 is a simple pole of t (cf., Proposition P. 7). This completes the proof.

2.2 COROLLARY. *Let* S(t) *be compact in* B. *Then for* $|\lambda| = 1$

$$t_n(\lambda) = \frac{1}{n}\left(\frac{t}{\lambda} + \frac{t^2}{\lambda^2} + \dots + \frac{t^n}{\lambda^n}\right) \to p_\lambda \quad (n \to +\infty)$$

in the norm of B. *Here* p_λ *is an idempotent in* B *such that*

$$(\lambda e - t)p_\lambda = p_\lambda(\lambda e - t) = 0.$$

Further $p_\lambda \neq 0$ *if and only if* $\lambda \in \sigma(t)$, *and in that case* λ *is a simple pole of* t.

PROOF. Observe that $|\lambda| = 1$ implies that

$$S(\tfrac{1}{\lambda}t) \subset \{\alpha x: |\alpha| = 1, x \in S(t)\}.$$

Clearly the latter set is compact, because $S(t)$ is compact. Since $S(\lambda^{-1}t)$ is closed, this implies that $S(\lambda^{-1}t)$ is compact. Hence the corollary is an immediate consequence of the preceding theorem.

Let $\mathbb{D}$ denote the closed unit disc and let Γ denote the unit circle in the complex plane $\mathbb{C}$.

2.3 THEOREM. *The semigroup* $S(t)$ *is compact in* B *if and only if* $\sigma(t) \subset \mathbb{D}$ *and* $\sigma(t) \cap \Gamma$ *is a finite (possibly empty) set of simple poles of of* t.

PROOF. Suppose that $S(t)$ is compact. Then $\{\|t^n\|\}$ is a bounded sequence. Hence

$$r(t) = \lim_{n\to+\infty} \|t^n\|^{\frac{1}{n}} \leq 1.$$

Therefore $\sigma(t) \subset \mathbb{D}$. If $\lambda \in \sigma(t) \cap \Gamma$, then, by Corollary 2.2, λ is a simple pole of t.

For the converse suppose first that $\sigma(t) \cap \Gamma$ is empty. Then $r(t) < 1$, and therefore (cf., Proposition P. 1)

$$t^n \to 0 \quad (n \to +\infty).$$

Hence $S(t)$ is compact. Alternatively, suppose that $\sigma(t) \cap \Gamma$ is a finite set of simple poles of t, $\lambda_1,\ldots,\lambda_k$ say, and let $p_1,\ldots,p_k$ be the associated spectral idempotents. Put

$$e_0 = p_1 + \ldots + p_k.$$

Then $r\{t(e - e_0)\} < 1$ (Proposition P. 4), and hence

$$t^n(e - e_0) = \{t(e - e_0)\}^n \to 0 \quad (n \to +\infty).$$

Further,

$$t^n e_0 = \lambda_1^n p_1 + \ldots + \lambda_k^n p_k.$$

Therefore $\{t^n e_0\}$ is a bounded sequence in a finite dimensional subspace of B. Combining these two facts, it follows that $S(t)$ is compact.

2.4 THEOREM. *Let* $S(t)$ *be a compact semigroup in* B. *Then* $\sigma(t) \cap \Gamma$ *is a spectral set of* t *and the spectral idempotent associated with this set is the unit of the kernel* K *of* $S(t)$. *Further, if* $\sigma(t) \cap \Gamma = \{\lambda_1,\ldots,\lambda_k\}$, *then*

K *is topologically isomorphic to the subgroup of the* k-*dimensional torus given by*

$$\mathrm{cl}\{(\lambda_1^n,\dots,\lambda_k^n)\colon n \in \mathbb{Z}^+\}.$$

PROOF. Let e_1 be the unit element of K. Then there exists an increasing sequence $\{n_i\}$ in $\mathbb{Z}^+$ such that

$$t^{n_i} \to e_1 \quad (i \to +\infty). \tag{5}$$

From this it follows that e_1 is a spectral idempotent associated with a spectral set σ of $\sigma(t)$ (Proposition P. 6). Further (5) implies that

$$t^{n_i}(e - e_1) \to 0 \quad (i \to +\infty).$$

Hence there exists $m \in \mathbb{Z}^+$ such that $r\{t^m(e - e_1)\} < 1$. But then $r\{t(e - e_1)\} < 1$. Hence (by Proposition P. 4)

$$\sigma \supset \sigma(t) \cap \Gamma. \tag{6}$$

From Theorem 2.3 we know that $\sigma(t) \cap \Gamma$ is a spectral set of t. Let e_0 be the associated spectral idempotent. Formula (6) implies that $e_1e_o = e_o$. On the other hand $r\{t(e - e_0)\} < 1$ by Theorem 2.3, so

$$t^n(e - e_0) = \{t(e - e_0)\}^n \to 0 \quad (n \to +\infty)$$

and thus

$$e_1 = \lim_i t^{n_i} = \lim_i t^{n_i} e_0 = e_1e_0.$$

Together these results show that $e_1 = e_0$.

Suppose $\sigma(t) \cap \Gamma = \{\lambda_1,\dots,\lambda_k\}$. Then λ_i is a simple pole of t. Let p_i be the associated spectral idempotent. Then

$$e_1 = p_1 + \dots + p_k$$

and

$$t^ne_1 = \lambda_1^np_1 + \dots + \lambda_k^np_k.$$

The map $\phi\colon \Lambda(te_1) \to \Gamma^k$ defined by

$$\phi(t^ne_1) = (\lambda_1^n,\dots,\lambda_k^n)$$

is a homeomorphism in the obvious topologies. Because of this, ϕ extends by continuity to a homeomorphism ψ between $K = S(te_1)$ and the semigroup

$$\mathrm{cl}\{(\lambda_1^n,\dots,\lambda_k^n)\colon n \in \mathbb{Z}^+\} \tag{7}$$

in Γ^k. Clearly, ψ is an isomorphism. So (7) is a subgroup of Γ^k topologically isomorphic to K. This completes the proof.

2.5 THEOREM. *The semigroup* $S(t)$ *is a compact group if and only if* $\sigma(t)$ *is a set of simple poles of* t *of modulus one or zero.*

PROOF. Suppose that $S(t)$ is a compact group. Let e_1 be its unit, then $t = te_1$ and so

$$t(e - e_1) = 0. \tag{8}$$

Since $S(t)$ is compact, we know from Theorem 2.3 that $\sigma(t) \cap \Gamma$ is a finite (possibly empty) set of simple poles of t. Further we can use Theorem 2.4 to show that $e - e_1$ is the spectral idempotent associated with the spectral set $\sigma(t) \setminus \Gamma$. But then, using (8) and Proposition P. 5, we see that either $\sigma(t) \setminus \Gamma$ is empty or $\sigma(t) \setminus \Gamma = \{0\}$, and then 0 is simple pole of t (by Proposition P. 7). Thus $\sigma(t)$ has the desired properties.

For the converse we note first of all that the hypothesis on $\sigma(t)$ implies that $S(t)$ is compact. Let e_1 be the unit of the kernel K of $S(t)$. According to Theorem 2.4, $e - e_1$ is the spectral idempotent associated with the spectral set $\sigma(t) \setminus \Gamma$. From what we know of $\sigma(t)$ this implies that $t(e - e_1) = 0$. Thus $t = te_1$ and hence

$$S(t) = S(te_1) = K$$

is a group.

3. SEPARATELY CONTINUOUS MULTIPLICATION

Let Ω be a topological semigroup in which multiplication is *separately continuous*, i.e., Ω is a semigroup endowed with a Hausdorff topology such that for each y in Ω the maps

$$x \to xy, \quad x \to yx$$

from $\Omega \to \Omega$ are continuous. If $t \in \Omega$, separate continuity ensures that

$$S(t) = \mathrm{cl}\{t^n : n \in \mathbb{Z}^+\}$$

is a commutative semigroup. Recall that the semigroup Ω is said to be monothetic if $\Omega = S(t)$ for some t in Ω. If $S(t)$ is compact, de Leeuw and Glicksberg [23] give information about its structure which can be considerably more complicated than in the jointly continuous case.

3.1 THEOREM. *Let* $S(t)$ *be a compact monothetic semigroup with separately continuous multiplication. Then*

(i) $K = \bigcap\{xS(t): x \in S(t)\}$ *is the unique minimal ideal of* $S(t)$;

(ii) K *is a compact topological group*;

(iii) *the unit* e_1 *of* K *is the unique minimal idempotent of* $S(t)$ (i.e., *for each idempotent* f *in* $S(t)$ *we have* $e_1 = fe_1$) *and*

$$K = e_1S(t) = S(e_1t).$$

PROOF. (i) The proof of Theorem 1.1(i) is valid for separately continuous multiplication.

(ii) The proof that K is algebraically a group which is a compact subset of $S(t)$ given in Theorem 1.1(ii) is valid for separately continuous multiplication. A deep result of Ellis [10] implies that multiplication in K is in fact jointly continuous, and now it follows as in Theorem 1.1(ii) that K is a topological group.

(iii) As in Theorem 1.1(iii) we prove that

$$K = e_1S(t) = S(e_1t). \tag{1}$$

Suppose that f is another idempotent in $S(t)$. From (1) we have

$$e_1f \in K.$$

But $S(t)$ is commutative, so e_1f is an idempotent in K. Therefore $e_1f = e_1$, that is, e_1 is the unique minimal idempotent in $S(t)$.

4. SEMIGROUPS OF OPERATORS IN THE WEAK OPERATOR TOPOLOGY

Let H be a Hilbert space, and let Ω denote the semigroup of all bounded linear operators on H endowed with the weak operator topology. Then (see [9], Exercises VI.9.11 and VI.9.6) multiplication in Ω is separately but not jointly continuous and the semigroup

$$\{T \in \Omega: \|T\| \leq 1\} \tag{1}$$

is compact in Ω.

Take T in Ω, and let $S(T)$ denote the closure in Ω of the set

$$\Lambda(T) = \{T^n: n = 1,2,\ldots\}.$$

Thus $S(T)$ is the monothetic semigroup in Ω generated by T. Note that $S(T)$ is compact in Ω if, and only if, the set $\Lambda(T)$ is bounded. For, if this is so, the set $\Lambda(T)$ and hence $S(T)$ is contained in some multiple of the set

(1). Conversely, if $S(T)$ is compact, then the set

$$\{\langle x,Ry\rangle : R \in S(T)\}$$

is bounded in $\mathbb{C}$ for each x and y in H. The uniform boundedness principle implies that $S(T)$ is norm bounded.

We shall say that x is a *univector* of a linear operator T if x is a (non-zero) eigenvector of T corresponding to a unimodular eigenvalue.

4.1 THEOREM. *Let* $S(T)$ *be a compact group in* Ω *with unit* I *the identity operator on* H. *Then* T *has a set of univectors in* H *whose closed linear span is* H.

PROOF. The information given about such a group in [15], section 22.23 implies that $S(T)$ is equivalent to a unitary group, that is, there exists a bijective element A in Ω such that

$$G = \{ARA^{-1} : R \in S(T)\}$$

is a group of unitary operators on H. Note that G is compact in Ω. The theory of unitary representations (see [15], section 22) shows that the operator ATA^{-1} has a complete orthonormal set of univectors in H. Whence the result.

Now let $S(T)$ be a compact semigroup in Ω, and let E_1 be the kernel idempotent of $S(T)$. If H_1 denotes the range of E_1 and H_2 its null space, we have the (not necessarily orthogonal) decomposition

$$H = H_1 \oplus H_2,$$

and since E_1 commutes with T we may write

$$T = T_1 \oplus T_2,$$

where T_i is the restriction of T to H_i $(i = 1,2)$. Observe that $\|T\| \leq 1$ implies that $\|E_1\| \leq 1$ and hence E_1 is orthogonal in that case.

4.2 THEOREM. *Let* $S(T)$ *be a compact semigroup in* Ω. *Then*

(i) *either* E_1 *is zero or* T_1 *has a set of univectors in* H_1 *whose closed linear span is* H_1;

(ii) $S(T_2)$ *contains the zero operator on* H_2.

PROOF. The kernel

$$E_1 S(T) = S(E_1 T)$$

is a compact group with unit E_1 (see Theorem 3.1). Hence $S(T_1)$ is a compact group with unit I_1. Here I_1 denotes the identity operator on H_1. So we can apply Theorem 4.1 to get (i).

To get (ii) observe that, since $E_1 \in S(T)$, there exists a net $\{n_\alpha\}$ in $\mathbf{Z}^+$ such that

$$T^{n_\alpha} \to E_1.$$

Therefore

$$T_2^{n_\alpha} \to 0_2,$$

where 0_2 denotes the zero operator on H_2.

Next we apply these results to a unitary operator U on H. Since $\|U^n\| = 1$ for $n \in \mathbf{Z}^+$, the semigroup S(U) is compact in Ω and its kernel idempotent is orthogonal.

4.3 THEOREM. *Let* U *be a unitary operator on* H. *Then* $\sigma(U)$ *is purely continuous spectrum if and only if*

$$0 \in S(U).$$

PROOF. Let E_1 be the kernel idempotent of S(U). Now the decomposition

$$H = H_1 \oplus H_2$$

associated with E_1 is orthogonal, thus the restrictions U_i of U to H_i (i = 1,2) are both unitary operators.

From Theorem 4.2(ii) we know that the zero operator 0_2 on H_2 belongs to $S(U_2)$. This implies that $\sigma(U_2)$ is purely continuous spectrum. For if $\sigma(U_2)$ contained a unimodular eigenvalue λ with corresponding univector x, we should have

$$|\langle U_2^n x, x\rangle| = |\langle \lambda^n x, x\rangle| = \langle x, x\rangle,$$

contradicting the fact that $0_2 \in S(U_2)$. On the other hand if H_1 is non-trivial, then $\sigma(U)$ does contain eigenvalues. Therefore $\sigma(U)$ is purely continuous spectrum if, and only if, $E_1 = 0$, and since E_1 is the minimal idempotent in S(U) (see Theorem 3.1(iii)), this last condition is equivalent to $0 \in S(U)$.

EXAMPLE. Let H be a Hilbert space with an orthonormal basis

$$\{e_n : n \in \mathbb{Z}\}.$$

Define the linear operator U on H by

$$Ue_n = e_{n+1} \qquad (n \in \mathbb{Z}).$$

The operator U is called the (unitary) *bilateral shift* on H. Observe that

$$\lim_n U^n = \lim_n U^{-n} = 0$$

in Ω. Thus $\sigma(U)$ is purely continuous spectrum. Further, we have

$$S(U) = \Lambda(U) \cup \{0\},$$

and the closure in Ω of the group $\{U^n : n \in \mathbb{Z}\}$ is no longer a group as it contains 0.

5. MANY-IDEMPOTENT SEMIGROUPS

We conclude this chapter with an example of a unitary operator generating a compact monothetic semigroup which contains uncountably many idempotents. The operator was originally constructed by West [33], who showed that the semigroup it generated contained the zero and the identity operators. This construction is given in Theorem 5.1. Brown and Moran [5] showed that this semigroup contained uncountably many idempotents and their work is contained in Theorem 5.2.

The construction of such an operator is based on the notion of a Kronecker set. Let Γ denote the circle group with the usual topology. A *Kronecker set* Δ of Γ is a subset of Γ such that any continuous function f on Δ of unit modulus can be uniformly approximated on Δ by characters of Γ, i.e., there is a sequence $\{n_i\}$ of integers such that

$$z^{n_i} \to f(z) \tag{1}$$

uniformly on Δ. From harmonic analysis (see [29], Chapter 5) one knows that there exists a Cantor set Δ in Γ which is a Kronecker set and which supports a non-zero continuous positive Borel measure μ. Let $L^2(\mu)$ denote the Hilbert space of μ-square integrable functions on Γ. Throughout the remainder of this section U will be the operator on $L^2(\mu)$ defined by

$$Uf(z) = zf(z)$$

for each f in $L^2(\mu)$. Then U is unitary and, as μ is a continuous measure,

$\sigma(U)$ is purely continuous spectrum.

5.1 THEOREM. *The monothetic semigroup* $S(U)$ *contains the zero and identity operators.*

PROOF. The operator U has purely continuous spectrum. Hence by Theorem 4.3, $0 \in S(U)$. Using the Kronecker property of Δ, choose a sequence $\{n_i\}$ in $\mathbb{Z}$ such that

$$z^{n_i} \to 1 \quad (i \to \infty) \tag{2}$$

uniformly on Δ. For $f,g \in L^2(\mu)$

$$\begin{aligned} |\langle (U^{n_i} - I)f,g\rangle| &= \left|\int_\Delta (z^{n_i} - 1)f(z)\overline{g(z)}d\mu(z)\right| \\ &\le \int_\Delta |z^{n_i} - 1||f(z)\overline{g(z)}|d\mu(z) \\ &\le \sup_{z\in\Delta} |z^{n_i} - 1| \int_\Delta |f(z)\overline{g(z)}|d\mu(z) \\ &\to 0 \quad (i \to \infty) \end{aligned}$$

because of (2). Hence

$$U^{n_i} \to I \quad (i \to \infty).$$

To see that $I \in S(U)$ we need to be able to choose the sequence $\{n_i\}$ in $\mathbb{Z}^+$. If $\{n_i\}$ has a subsequence in $\mathbb{Z}^+$, then it will do. If not, $\{n_i\}$ is ultimately in $\mathbb{Z}^-$. Put $m_i = -n_i$ for each i and use the fact that the adjoint operation is continuous in the weak operator topology ([9], Exercise VI.9.12) to get

$$U^{m_i} = U^{n_i *} \to I \quad (i \to \infty).$$

Thus $I \in S(U)$, concluding the proof.

5.2 THEOREM. *The monothetic semigroup* $S(U)$ *contains uncountably many idempotents.*

PROOF. Take $\delta \in \Delta$, and define the function v on Δ to be the restriction to Δ of the function

$$w(z) = \begin{cases} \overline{\delta}z & \text{if } 0 < \arg z \le \arg \delta, \\ 1 & \text{otherwise.} \end{cases}$$

Since $z = 1$ cannot belong to any Kronecker set, v is continuous on Δ. Further v is unimodular. Hence there exists a sequence $\{n_i\}$ in Z such that

$$z^{n_i} \to v \quad (i \to \infty)$$

uniformly on Δ. Define the operator V on $L^2(\mu)$ to be multiplication by v. Then V is a unitary operator and, as in the proof of Theorem 5.1, it follows that either V or V^* (multiplication by the conjugate of v) or both lie in $S(U)$. We shall suppose that the sequence $\{n_i\}$ has been chosen in $\mathbf{Z}^+$ and, therefore, $V \in S(U)$. (The proof for V^* requires only obvious changes of detail).

Now Δ is the union of two sets, namely $\Delta_1 = \{z \in \Delta: v(z) = 1\}$ and $\Delta_2 = \{z \in \Delta: v(z) = \bar{\delta}z\}$. Both are Borel sets and their intersection has μ-measure zero, thus every f in $L^2(\mu)$ can be uniquely decomposed into the sum of its restrictions to Δ_1 and Δ_2. This induces an orthogonal decomposition of $L^2(\mu)$ into

$$L^2(\mu) = H_1 \oplus H_2.$$

Observe that H_1 and H_2 are invariant under U and therefore under V. Further the restriction U_i of U to H_i is unitary for $i = 1,2$, and

$$V = I_1 \oplus \bar{\delta}U_2,$$

where I_1 is the identity operator on H_1. The spectrum of U and hence of its restriction U_2 is purely continuous spectrum, therefore, by Theorem 4.3, the zero operator O_2 on H_2 belongs to $S(U_2)$. The Hilbert spaces involved are all separable, so the weak operator topology restricted to the unit ball of operators is metrizable. Thus there exists a sequence $\{n_k\}$ in $\mathbf{Z}^+$ such that

$$U_2^{n_k} \to O_2 \quad (k \to +\infty).$$

Therefore

$$V^{n_k} \to I_1 \oplus O_2 = P \quad (k \to +\infty),$$

where P is the idempotent corresponding to multiplication by the characteristic function of Δ_1. Since $V \in S(U)$, we have $P \in S(U)$, and we can construct an uncountable number of distinct such idempotents by varying δ and using the fact that μ is continuous.

Brown and Moran have carried their investigations considerably further. In [5] they give examples of measures μ on Γ such that the corresponding operator semigroup $S(U)$ has (i) a countable infinity of idempotents or (ii) any prescribed finite non-zero number of idempotents. In cases (i) and (ii) the idempotents are totally ordered. In [6] they construct a measure μ such that S(U) contains just four idempotents which are not totally ordered. They round off this discussion in [7] by showing that any finite lower semilattice can be realised as the idempotent subsemigroup of S(U) for a suitable μ. Their methods are based on some extremely intricate harmonic analysis of the unit circle.

NOTES ON CHAPTER I

1. Let S be a finite semigroup with n distinct elements generated by a single element t, i.e.,

$$S = \{t, t^2, \ldots, t^n\}.$$

The structure of such a cyclic semigroup is as follows. Let ℓ be the unique positive integer such that $\ell \leq n$ and

$$t^{n+1} = t^{\ell}.$$

Then the kernel of S is the set

$$K = \{t^{\ell}, t^{\ell+1}, \ldots, t^n\}$$

and t^{ℓ} is the kernel idempotent of S (see [25], p.110).

2. Theorem 1.1(i) and (ii) are valid for any compact commutative semigroup with jointly continuous multiplication.

3. In addition to Theorem 1.1 we have the following results valid under the same hypothesis (see [15], section 9.28).

(a) $K = \bigcap_{k=1}^{\infty} \mathrm{cl}\{t^n : n \geq k\}$;

(b) $S(t) = \Lambda(t) \cup K$;

(c) if S(t) has two distinct generators, it is a group.

Observe that (a) is equivalent to the statement that K is the set of cluster points of the sequence $\{t, t^2, \ldots\}$ (cf. the last paragraph of section 1). The semigroup S(U), employed in Theorem 5.1, shows that results (a), (b) and (c) all fail in the case of separately continuous multiplication.

4. Theorems 2.3 and 2.5 could be proved directly from Theorem 1.1 and the representation theory of uniformly compact groups. However the ergodic approach is important in later discussions of semi-algebras (see section

III.2).

5. Kaashoek and West [21] have investigated the connection between the structure of a compact commutative semigroup in a Banach algebra and the spectrum (Gelfand space) of the closed subalgebra it generates.

6. Theorem 3.1 (with some obvious modifications) is valid for any compact commutative semigroup with separately continuous multiplication.

7. From the work of the Leeuw and Glicksberg [23], one can see that Theorems 4.1 and 4.2 hold for operators T on a Banach space.

8. In [14] Gillespie and West make a detailed study of operators on a Banach space which generate weakly compact groups. In Theorem 1.2 of that paper they give the following converse of our Theorem 4.1. Let T be a bounded linear operator on a Banach space X such that

$$\|T^n\| \leq M \quad (n \in \mathbb{Z}^+),$$

and let T have a set of univectors whose closed linear span is X, then the monothetic semigroup generated by T in the weak operator topology is a compact group with unit the identity operator on X. This result is due to Kaashoek and Wolff.

CHAPTER II

LOCALLY COMPACT SEMI-ALGEBRAS

The first main result of this chapter (Theorem 1.1) gives general sufficient conditions for a non-zero element in a Banach algebra to generate a locally compact semi-algebra. This generalizes earlier results of Bonsall and Tomiuk [4] and the authors [19]. The conditions may be loosely stated as follows: Apart from a finite set of poles which do not lie on the positive real axis the peripheral spectrum of the generator consists of a set of poles, one of which is the spectral radius. This result is due to Kaashoek (see Chapter I of [1]).

A semi-algebra is said to be strict if the sum of any two elements in the semi-algebra is zero only if both of them are zero. Theorem 1.2 (also due to Kaashoek [17]) gives sufficient conditions for a non-zero element of a Banach algebra to generate a strict, locally compact semi-algebra. The conditions are that the peripheral spectrum of the generator should consist of poles and that among these the spectral radius should be of maximal order. The proofs of Theorems 1.1 and 1.2 are long but use only elementary spectral theory.

Sections 2 and 3 of Chapter II are based on Bonsall's work [2] on the structure of locally compact semi-algebras, although in some cases the proofs are different. Local compactness implies the existence of minimal closed two-sided ideals (Lemma 2.1) and this, in turn, leads to the existence of idempotents (Theorem 2.5). One obtains a theory closely related to classical Wedderburn theory; the final result (Theorem 3.1) being a characterization of the semi-algebra of positive matrices of fixed order as a strict locally compact semi-algebra which possesses no proper closed two-sided ideals.

1. DEFINITIONS AND EXAMPLES

Let B be a complex Banach algebra with unit e. A non-empty subset A of B is said to be a *semi-algebra* if A is closed under addition, multiplication and multiplication by non-negative real scalars, i.e.,

$$a,b \in A, \quad \alpha \geq 0 \Rightarrow a + b,\ ab,\ \alpha a \in A.$$

A semi-algebra A is called *locally compact* if A contains non-zero elements and

$$A \cap \{x \in B: \|x\| \leq 1\}$$

is a compact subset of B. These axioms imply that A is a closed subset of B and that A is a locally compact topological space with respect to the relative topology induced in A by the norm topology of B.

A semi-algebra A is called *closed* if A is a closed subset of B. The intersection of a set of closed semi-algebras in B is a closed semi-algebra. Thus, if $t \in B$, the smallest closed semi-algebra containing t exists. This set is denoted by A(t). Note that A(t) is the closure in B of the set

$$\{\alpha_1 t + \ldots + \alpha_k t^k: \alpha_i \geq 0\ (i = 1,\ldots,k),\ k \in \mathbb{Z}^+\}.$$

A semi-algebra A is called *monothetic* if A = A(t) for some t in A. In that case we say that A is *generated by* t, and we call t a *generator* of A.

A semi-algebra A is *commutative* if

$$ab = ba \quad (a,b \in A).$$

Obviously monothetic semi-algebras are commutative.

Let A be a semi-algebra in B. The set

$$B_o = A - A = \{a - b: a,b \in A\}$$

is a real subalgebra of B. If A contains non-zero elements and B_o is finite dimensional, then A is trivially locally compact. The next theorem shows that we can construct examples which are not of this type.

1.1 THEOREM. *Let* t *be a non-zero element in* B. *Suppose that* $\sigma(t)$ *decomposes into two disjoint closed subsets* σ_1 *and* σ_2 *such that*

(i) σ_1 *is a finite (possibly empty) set of poles of* t *and*

$\sigma_1 \cap \mathbb{R}^+ = \emptyset$;

(ii) *either* $\sigma_2 = \emptyset$ *or there exists* $0 < \alpha \in \sigma_2$ *such that*

$$\sigma_2 = \sigma(t) \cap \{\lambda: |\lambda| \leq \alpha\}$$

and

$$\sigma_2 \cap \{\lambda: |\lambda| = \alpha\}$$

is a finite set of poles of t.

Then A(t) *is locally compact.*

PROOF. The sets σ_1 and σ_2 are spectral sets of t. Let p_i be the spectral idempotent associated with σ_i (i = 1,2). Then

$$e = p_1 + p_2, \quad p_1p_2 = 0.$$

For any a in A(t), $ap_i \in A(tp_i)$ for i = 1,2 and

$$a = ap_1 + ap_2.$$

Hence it suffices to show that, for i = 1,2, either $A(tp_i)$ is locally compact or consists of the zero element only.

Since σ_1 is a finite (possibly empty) set of poles of t, the closed subalgebra of B generated by tp_1 is finite dimensional (Proposition P.8). Hence $A(tp_1)$ lies in a finite dimensional subalgebra of B. This implies that either $A(tp_1)$ is locally compact or $A(tp_1)$ consists of the zero element only.

If $\sigma_2 = \emptyset$, then $A(tp_2) = \{0\}$. Suppose $\sigma_2 \neq \emptyset$. We shall prove that $t_2 = tp_2$ generates a locally compact semi-algebra. First of all note that $r(t_2) = \alpha$, because

$$\sigma_2 \subset \sigma(t_2) \subset \sigma_2 \cup \{0\}$$

and $\alpha \in \sigma_2$. Further

$$\sigma(t_2) \cap \{\lambda: |\lambda| = \alpha\}$$

is a finite set of poles of t and thus of t_2. Replacing t_2 by $\alpha^{-1}t_2$ we see that without loss of generality we may assume that $r(t_2) = \alpha = 1$.

Take b_n in $A(t_2)$, and assume that $\|b_n\| \leq 1$ for n = 1,2,... . We have to prove that the sequence $\{b_n\}$ has a convergent subsequence. Since b_n is in the closure of the set

$$\{\alpha_1t_2 + \ldots + \alpha_kt_2^k: \alpha_i \geq 0, k \in \mathbf{Z}^+\},$$

it suffices to verify this statement for the case that

$$b_n = \sum_{k=1}^{\infty} \alpha_k(n) t_2^k,$$

where $\alpha_k(n) \geq 0$ and $\alpha_k(n) = 0$ for k sufficiently large. Since $r(t_2) = 1 \in \sigma(t_2)$, the spectral mapping theorem (Proposition P.3) implies that

$$r(b_n) = \sum_{k=1}^{\infty} \alpha_k(n).$$

Now $r(b_n) \leq \|b_n\| \leq 1$. Hence

$$\sum_{k=1}^{\infty} \alpha_k(n) \leq 1.$$

By using compactness arguments and the diagonal process, we obtain an increasing sequence $\{n_i\}$ in $\mathbf{Z}^+$ such that for each k the sequence $\{\alpha_k(n_i)\}$ converges. By passing to this subsequence, we may suppose that

$$\beta_k = \lim_n \alpha_k(n)$$

exists. Observe that

$$\beta_k \geq 0, \quad \sum_{k=1}^{\infty} \beta_k \leq 1.$$

Let q be the spectral idempotent associated with the spectral set

$$\sigma(t_2) \cap \{\lambda\colon |\lambda| = 1\},$$

and put $s = t_2(e - q)$. Since $r(s) < 1$, we know that $\|s^n\| \to 0$ for $n \to +\infty$. This implies that the sequence

$$\sum_{k=1}^{\infty} \beta_k s^k$$

converges in the norm of B. Let d be its sum. Then the foregoing shows that

$$d = \lim_n b_n(e - q).$$

Now $b_n = b_n(e - q) + b_n q$ for each n in $\mathbf{Z}^+$. Hence in order to prove that $\{b_n\}$ has a convergent subsequence, it suffices to show that $\{b_n q\}$ has a convergent subsequence. But $\{b_n q\}$ is a bounded sequence in the closed subalgebra of B generated by $t_2 q$. From the spectral properties of t_2 it follows that this algebra is finite dimensional (cf., Proposition P.8). So $\{b_n q\}$ has a convergent subsequence, and the proof is complete.

Let t be a non-zero element in B satisfying the conditions of the previous theorem and suppose that $\sigma(t)$ is an infinite set. Then $B_o = A(t) - A(t)$ is not finite dimensional. For suppose that B_o is finite dimensional. Since $t^n \in B_o$ for each n in $\mathbf{Z}^+$, this implies the existence of a non-zero polynomial q such that $q(t) = 0$. But then $\sigma(t)$ must be finite (by Proposition P.3), contradicting our hypothesis.

Using Theorem 1.1, it is easy to construct examples of monothetic locally compact semi-algebras. We mention one particular one.

EXAMPLE. Let T be a compact linear operator on a complex Banach space E, and suppose that T has a positive eigenvalue. Then the monothetic semi-algebra generated by T in the Banach algebra L(E) of all bounded linear operators on E is locally compact.

A semi-algebra A in B is called *strict* if

$$A \cap (-A) = \{0\},$$

i.e., if -x and x in A implies x = 0. The next theorem shows how one may construct strict locally compact monothetic semi-algebras. The result is formulated in terms of the set

$$\text{Per}\sigma(t) = \{\lambda \in \sigma(t) : |\lambda| = r(t)\}.$$

1.2 THEOREM. *Let* t *be a non-zero element of* B *such that* Perσ(t) *is a set of poles of* t. *Further suppose that* r(t) *is a pole of* t *of maximal order in* Perσ(t). *Then* A(t) *is strict and locally compact.*

PROOF. The proof consists of two parts.

1. First of all we consider the case r(t) > 0. The local compactness of A(t) follows immediately from Theorem 1.1 by taking $\sigma_1 = \emptyset$ and $\sigma_2 = \sigma(t)$. Let $s = \beta t$ for some $\beta > 0$. Then A(s) = A(t), $r(s) = \beta r(t)$, and s and t have similar spectral properties. Hence in order to prove that A(t) is strict, we may suppose without loss of generality that r(t) = 1. Then 1 is a pole of t of order n_0, say. Let p_0 be the spectral idempotent associated with the spectral set {1}. Put $x_0 = (e - t)^{n_0 - 1} p_0$. Then (by Proposition P.7)

$$x_0 \neq 0, \quad tx_0 = x_0.$$

Let $b(t) = \alpha_1 t + \ldots + \alpha_n t^n$ with $\alpha_i \geq 0$ $(i = 1,\ldots,n)$. The spectral mapping theorem (Proposition P.3) implies that

$$r(b(t)) = b(1).$$

Further we have

$$b(t)x_0 = b(1)x_0 = r(b(t))x_0.$$

Take s in A(t) ∩ (-A(t)). We want to show that s = 0. Since s ∈ A(t), there exists a sequence $\{b_n\}$ in $A(t)$ such that $s = \lim b_n$ and

$$b_n = \sum_{i=1}^{\infty} \alpha_i(n)t^i \quad (n = 1,2,\dots)$$

with $\alpha_i(n) \geq 0$ and $\alpha_i(n) = 0$ for i sufficiently large. From what we proved in the previous paragraph it follows that

$$r(b_n) = \sum_{i=1}^{\infty} \alpha_i(n), \quad b_n x_o = r(b_n)x_o$$

for $n = 1,2,\dots$. By the continuity of the spectral radius on commutative sets (Proposition P.2), the last equality implies that

$$sx_o = r(s)x_o. \tag{1}$$

Since $-s$ also belongs to $A(t)$, we can repeat the argument to show that

$$-sx_o = r(-s)x_o. \tag{2}$$

But $r(-s) = r(s)$. Hence (1) and (2) together imply that $r(s) = 0$, and so

$$\lim_n \sum_{i=1}^{\infty} \alpha_i(n) = \lim_n r(b_n) = r(s) = 0. \tag{3}$$

From our hypothesis it follows that the peripheral spectrum of t is a spectral set of t. Let p be the corresponding spectral idempotent. Then $r\{t(e - p)\} < 1$, and so

$$t^n(e - p) = \{t(e - p)\}^n \to 0 \quad (n \to +\infty).$$

In particular this implies that the sequence $\{t^n(e - p)\}$ is bounded in B, and so by (3)

$$\lim_n \sum_{i=1}^{\infty} \alpha_i(n)t^i(e - p) = 0.$$

That is, $s(e - p) = \lim b_n(e - p) = 0$. Hence in order to prove that $s = 0$, it is sufficient to show that $sp = 0$.

Let

$$\mathrm{Per}\sigma(t) = \{\lambda_o=1,\lambda_1,\dots,\lambda_r\}.$$

Then $|\lambda_i| = 1$, λ_i is a pole of t of order n_i say, and $n_i \leq n_o$ ($i = 0,1,2,\dots$ $\dots, r$). Let p_i be the spectral idempotent associated with the spectral set $\{\lambda_i\}$. Then

$$p = p_o + p_1 + \dots + p_r.$$

Hence it suffices to show that $sp_i = 0$ for $i = 0,1,\dots,r$.

Put $\alpha_o(n) = 0$ for each n in $\mathbb{Z}^+$, and let

$$\beta_j^{\ell}(n) = \sum_{i=j}^{\infty} \alpha_i(n)\binom{i}{j}\lambda_{\ell}^{i-j}$$

for $\ell = 0,1,\dots,r$ and $j = 0,1,\dots$ and each n in $\mathbb{Z}^+$. Observe that for fixed n and ℓ and for j sufficiently large $\beta_j^{\ell}(n) = 0$. Consider the complex polynomial

$$q_n^{\ell}(X) = \sum_{j=0}^{\infty} \beta_j^{\ell}(n)X^j$$

for $\ell = 0,1,\dots,r$ and each n in $\mathbf{Z}^+$. A simple computation shows that

$$q_n^{\ell}(X) = \sum_{i=0}^{\infty} \alpha_i(n)(\lambda_{\ell} + X)^i.$$

In particular

$$b_n = q_n^{\ell}(t - \lambda_{\ell}) \tag{4}$$

for $\ell = 0,1,\dots,r$ and each n in $\mathbf{Z}^+$. Note that

$$|\beta_j^{\ell}(n)| \leq \sum_{i=j}^{\infty} \alpha_i(n)\binom{i}{j}|\lambda_{\ell}|^{i-j} = \beta_j^{o}(n) \tag{5}$$

for $\ell = 0,1,\dots,r$ and $j = 0,1,2,\dots$ and each n in $\mathbf{Z}^+$.

Since $\lambda_o = 1$ is a pole of t of order n_o, the elements

$$p_o,\ (t - \lambda_o)p_o,\ \dots\ ,\ (t - \lambda_o)^{n_o-1}p_o \tag{6}$$

are linearly independent. Further $(t - \lambda_o)^{n_o}p_o = 0$, and thus by (4)

$$b_np_o = q_n^{o}(t - \lambda_o)p_o = \sum_{j=0}^{n_o-1} \beta_j^{o}(n)(t - \lambda_o)^j p_o.$$

Now $sp_o = \lim b_np_o$. Hence

$$sp_o = \sum_{j=0}^{n_o-1} \beta_j^{o}(\infty)(t - \lambda_o)^j p_o, \tag{7}$$

where

$$\beta_j^{o}(\infty) = \lim_n \beta_j^{o}(n) \quad (j = 0,1,\dots,n_o-1).$$

Note that (5) implies that $\beta_j^{o}(\infty) \geq 0$ for $j = 0,1,\dots,n_o-1$.

Also -s in A(t). So repeating the argument we see that

$$-sp_o = \sum_{j=0}^{n_o-1} \gamma_j^{o}(t - \lambda_o)^j p_o \tag{8}$$

for some $\gamma_j^{o} \geq 0$ $(j = 0,1,\dots,n_o-1)$. But the elements (6) are linearly independent, therefore (7) and (8) together imply that $\beta_j^{o}(\infty) = 0$ for $j = 0,1,\dots,n_o-1$, and hence by (5)

$$\lim_n \beta_j^{\ell}(n) = 0 \tag{9}$$

for $\ell = 0,1,\dots,r$ and $j = 0,1,\dots,n_o-1$.

Since λ_{ℓ} is a pole of t of order n_{ℓ} with $n_{\ell} \leq n_o$, we have $(t - \lambda_{\ell})^{n_o}p_{\ell} = 0$. Hence (4) implies that

$$b_np_{\ell} = q_n^{\ell}(t - \lambda_{\ell})p_{\ell} = \sum_{j=0}^{n_o-1} \beta_j^{\ell}(n)(t - \lambda_{\ell})^j p_{\ell}$$

for $\ell = 0,1,\dots,r$ and each n in Z^+. But then we can use (9) to show that

$$sp_\ell = \lim_n b_n p_\ell = 0 \quad (\ell = 0,1,\dots,r).$$

This completes the proof of part 1.

2. Next we consider the case $r(t) = 0$. Then $\sigma(t) = \{0\}$, and the local compactness of A(t) follows immediately from Theorem 1.1 by taking $\sigma_1 = \sigma(t)$ and $\sigma_2 = \emptyset$. Observe that the assumption $r(t) = 0$ together with the hypotheses of the theorem implies that t is nilpotent. Let n_o be the order of nilpotence. If $n_o = 1$, then $t = 0$, contradicting the hypotheses of the theorem. Thus $n_o \geq 2$.

Take s in $A(t) \cap (-A(t))$. We want to show that $s = 0$. Let the sequence $\{b_n\}$ in A(t) be as in part 1. Then

$$b_n = \sum_{i=1}^{n_o-1} \alpha_i(n)t^i \quad (n = 1,2,\dots).$$

Using the linear independence of the elements $t, t^2,\dots,t^{n_o-1}$ we see that

$$s = \sum_{i=1}^{n_o-1} \alpha_i(\infty)t^i, \tag{10}$$

where $\alpha_i(\infty) = \lim \alpha_i(n) \geq 0$ for $i = 1,\dots,n_o-1$. Also $-s$ in A(t). So we can repeat the argument to show that

$$-s = \sum_{i=1}^{n_o-1} \delta_i t^i \tag{11}$$

for some $\delta_i \geq 0$ $(i = 1,\dots,n_o-1)$. Since the elements $t, t^2,\dots,t^{n_o-1}$ are linearly independent, (10) and (11) together imply that $\alpha_i(\infty) = 0$ for $i = 1,\dots,n_o-1$. Thus $s = 0$. This completes the proof.

One of our aims is to get converses of the preceding two theorems. For Theorem 1.1 this will be the main topic of Chapter III. The results of Chapter III are based on the existence of idempotents in locally compact semi-algebras, a subject we will be dealing with in the next section.

2. IDEMPOTENTS

In this section B denotes a complex Banach algebra with unit element e. Let A be a semi-algebra in B. A non-empty subset J of A is called a *right ideal* of A if

(i) J is a semi-algebra,

(ii) $x \in A$, $a \in J$ implies $ax \in J$.

Similarly one defines a *left ideal*. The set J is said to be a *two-sided ideal* if J is both a left ideal and a right ideal. An ideal (left, right or two-sided) J of A is called a *closed* ideal if J is a closed subset of A with respect to the relative topology induced in A by the norm topology of B. Note that closed ideals in a locally compact semi-algebra are closed subsets of B.

A closed right ideal J of A is called a *minimal closed right ideal* if $J \neq (0)$ and the only closed right ideals of A contained in J are (0) and J. Similar definitions apply for minimal closed left or two-sided ideals.

2.1 LEMMA. *In a locally compact semi-algebra each non-zero closed right ideal contains a minimal closed right ideal. A similar statement holds for left and two-sided ideals.*

PROOF. Let J be a non-zero closed right ideal of the locally compact semi-algebra A. Further let W be the family of all non-zero closed ideals of A contained in J. The set W is non-empty, and W is partially ordered by the inclusion relation $\subset$. The ordering on W is inductive. To see this, let V be a non-empty linearly ordered subset of W. Put

$$V_1 = \{I \cap \{x \in B: \|x\| = 1\}: I \in V\}.$$

Then V_1 is a family of compact sets and V_1 has the finite intersection property. Hence $\cap V_1 \neq \emptyset$ and thus $J_o = \cap V$ contains non-zero elements. Note that J_o is a closed right ideal of A contained in J. Hence J_o is a lowerbound for V in W. So W is inductively ordered, and we can apply Zorn's lemma to show the existence of a minimal closed right ideal contained in J.

The "minimum condition" on the closed ideals proved above is the main tool in the structure theory of locally compact semi-algebras. In this section we develop this theory as far as is necessary to get suitable conditions ensuring the existence of idempotents.

Let D be a non-empty subset of the semi-algebra A. The *right annihilator* D_r of D is the set of all x in A such that ax = 0 for each a in D. Thus

$$D_r = \{x \in A: ax = 0 \quad (a \in D)\}.$$

A *left annihilator* is defined similarly.

Observe that D_r is a closed right ideal of A. If D itself is a right ideal, then it is easily seen that D_r is a closed two-sided ideal.

2.2 LEMMA. *Let* E *be a closed subset of the locally compact semi-algebra* A *and suppose that* $\alpha E \subset E$ $(\alpha \geq 0)$. *Let* a *be an element in* A *such that*

$$\{a\}_r \cap E = \{0\}. \tag{1}$$

Then aE *is closed.*

PROOF. Let $y = \lim ax_n$, where $\{x_n\}$ is a sequence in E. Firstly, suppose that the sequence $\{x_n\}$ is bounded. The local compactness of A implies the existence of a convergent subsequence $\{x_{n_i}\}$ with limit x_o, say. Then $y = ax_o$, and $x_o \in E$, since E is closed. Hence $y \in aE$, which is the desired result.

Next suppose $\{x_n\}$ to be unbounded. Then there exists a subsequence $\{x_{n_i}\}$ such that $\|x_{n_i}\| \geq i$ for $i = 1,2,\ldots$. Put

$$x_i' = \|x_{n_i}\|^{-1} x_{n_i} \quad (i = 1,2,\ldots).$$

Note that $\lim ax_i' = 0$. The hypothesis on E implies that $x_i' \in E$ for each i. Since $\|x_i'\| = 1$, we can repeat the argument of the first part of the proof to show that $\{x_i'\}$ has a limit point x_o' in E and $ax_o' = 0$. By (1), $x_o' = 0$. However, this contradicts the fact that $\|x_i'\| = 1$ for each i. Hence the sequence $\{x_n\}$ is not unbounded.

An idempotent p in A is called a *right minimal idempotent* of A if pA is a minimal closed right ideal. Note that minimal idempotents are non-zero.

2.3 PROPOSITION. *Let* A *be a locally compact semi-algebra, and let* M *be a minimal closed right ideal in* A. *Suppose that* p *is a non-zero idempotent in* M. *Then*

$$pm = m \quad (m \in M).$$

In particular pA = M, *and thus* p *is a right minimal idempotent.*

PROOF. Put $J = \{m \in M: m - pm \in M\}$. Then J is a closed right ideal contained in M. Since $0 \neq p \in J$ and M is minimal, we have J = M. Each element $m - pm \in \{p\}_r$. Hence the foregoing implies that

$$m - pm \in \{p\}_r \cap M \quad (m \in M).$$

The set $\{p\}_r \cap M$ is a closed right ideal contained in M. Since $p \notin \{p\}_r$,

$\{p\}_r \cap M \neq M$. Hence $\{p\}_r \cap M = (0)$, and thus $m = pm$ for each $m \in M$. The remaining assertions are trivial consequences of this result.

2.4 PROPOSITION. *Let* A *be a locally compact semi-algebra, and let* M *be a minimal closed right ideal in* A. *Take* a *in* A *and suppose that* $aM \neq (0)$. *Then* aM *is a minimal closed right ideal.*

PROOF. The set $\{a\}_r \cap M$ is a closed right ideal of A contained in M. From our hypotheses it follows that $\{a\}_r \cap M \neq M$. Since M is minimal, this implies that

$$\{a\}_r \cap M = (0).$$

So we can apply Lemma 2.2 to show that aM is closed.

Clearly aM is a right ideal. It remains to show that aM is minimal. Let I be a closed right ideal contained in aM, and suppose that $I \neq aM$. Let $J = \{m \in M: am \in I\}$. The set J is a closed right ideal contained in M and $J \neq M$. Hence $J = (0)$, and thus $I = (0)$. This shows that aM is a minimal closed right ideal.

2.5 THEOREM. *Let* A *be a locally compact semi-algebra, and let* M *be a minimal closed right ideal in* A. *Suppose that* $M^2 \neq (0)$. *Then* M *contains a non-zero idempotent.*

PROOF. The proof consists of four parts.

1. From our hypotheses follows the existence of an element t in M such that $tM \neq (0)$. Let $J = \{t\}_r \cap M$. Then J is a closed right ideal contained in M and $J \neq M$. Since M is minimal this implies that $J = (0)$. Hence

$$tx \neq 0 \quad (0 \neq x \in M). \tag{2}$$

2. Assume that A is commutative. From (2) it follows that $xM \neq (0)$ for each non-zero x in M, and hence we can apply Proposition 2.4 to show that xM is a closed right ideal. Further $xM \subset M$ $(x \in M)$. So

$$xM = M \quad (0 \neq x \in M).$$

But this implies that $M \setminus \{0\}$ is a group with respect to multiplication. The unit element of this group is a non-zero idempotent contained in M. This completes the proof for the commutative case.

3. Next we return to the general case. Let t be as in part 1. Consider A(t), the monothetic semi-algebra generated by t. Since $A(t) \subset M$, (2)

implies that $ts \neq 0$ for each non-zero s in A(t). But A(t) is commutative. Thus $st \neq 0$ for $0 \neq s \in A(t)$. Hence

$$sM \neq (0) \quad (0 \neq s \in A(t)).$$

Repeating the arguments of part 1 for s instead of t we see that

$$sx \neq 0 \quad (0 \neq x \in M).$$

In particular $s_1 s_2 \neq 0$ for each pair s_1 and s_2 of non-zero elements in A(t).

4. Observe that A(t) is a commutative locally compact semi-algebra. According to Lemma 2.1, A(t) contains a minimal closed ideal N, say. Since A(t) has no divisors of zero (see the conclusion of part 3), we have $N^2 \neq (0)$. But then, by the result of part 2, N contains a non-zero idempotent p. Since

$$N \subset A(t) \subset M,$$

p is a non-zero idempotent in M.

An element p in a semi-algebra A is called a *unit element* of A if $p \neq 0$ and

$$pa = ap = a \quad (a \in A).$$

A semi-algebra A is called a *division semi-algebra* if A has a unit element p and each non-zero element x of A has an inverse y in A, i.e.

$$xy = yx = p.$$

Equivalently, A is a division semi-algebra if

$$A^* = A \setminus \{0\}$$

is a group with respect to multiplication.

2.6 THEOREM. *Let* p *be a right minimal idempotent in the locally compact semi-algebra* A. *Then* pAp *is a locally compact division semi-algebra.*

PROOF. Let D = pAp. Then D is clearly a semi-algebra with unit element p. Further D is a closed subset of A, since

$$D = \{x \in A: x = px = xp\}.$$

The local compactness of A implies that D is locally compact.

It remains to prove that $D^* = D \setminus \{0\}$ is a group with respect to multiplication. Let pap be in D^*. We shall prove that the equation

$$(pap)X = p \tag{3}$$

has a solution in D^*. To show this, it suffices to prove that the equation

$$(pa)Y = p \tag{4}$$

is solvable in pA.

From our hypothesis it follows that M = pA is a minimal closed right ideal of A. Since $p^2 = p$,

$$0 \neq pap \in (pa)M.$$

So we can apply Proposition 2.4 to show that (pa)M is a closed right ideal. But

$$(pa)M \subset (pA)M = M^2 \subset M.$$

Since M is minimal, this implies (pa)M = M. Hence equation (4) has a solution in M = pA.

The fact that equation (3) is solvable in D^* proves that each element in D^* has a right inverse in D^*. Using a standard argument from elementary group theory, one can show that this implies that D^* is closed under multiplication. Hence it follows that D is a group with unit element p.

2.7 THEOREM. *Let* A *be a strict closed division semi-algebra with unit element* p. *Then*

$$A = \mathbb{R}_o^+ p.$$

PROOF. Let $u \in A$. First of all we show that $\lambda p - u \in A$ for $\lambda > \|u\|$. Consider the series

$$\sum_{n=1}^{\infty} \frac{u^{n-1}}{\lambda^n}$$

with $u^o = p$. For $\lambda > \|u\|$ this series converges in B. Let b_λ denote its sum. Since the partial sums of the series belong to A and A is closed, $b_\lambda \in A$. Further we have $b_\lambda(\lambda p - u) = (\lambda p - u)b_\lambda = p$. Hence

$$\lambda p - u = b_\lambda^{-1} \in A.$$

Here x^{-1} denotes the inverse of x in A.

Let $\mu = \inf\{\lambda\colon \lambda p - u \in A\}$. Since A is strict, $\mu \geq 0$. Since A is closed, $\mu p - u \in A$. We shall prove that

$$y = \mu p - u = 0.$$

Suppose not. Then, by the argument of the first paragraph, $\nu p - y^{-1} \in A$ for ν sufficiently large and positive. But then

$$(\mu - \varepsilon)p - u = y - \varepsilon p = \varepsilon y(\tfrac{1}{\varepsilon}p - y^{-1}) \in A$$

for $\varepsilon > 0$ sufficiently small. However this contradicts the definition of μ. Hence $y = 0$ and $u = \mu p$. This shows that $A \subset \mathbb{R}_o^+ p$. Since $p \in A$ the reverse inclusion holds trivially. Thus $A = \mathbb{R}_o^+ p$.

3. SIMPLE LOCALLY COMPACT SEMI-ALGEBRAS

A semi-algebra A is said to be *simple* if (0) and A are the only two-sided ideals of A. Clearly a division semi-algebra is simple. The n×n matrices with non-negative entries provide a less trivial example.

Let $M_n(\mathbb{C})$ be the algebra of all n×n matrices with complex entries endowed with some algebra norm. With such a norm $M_n(\mathbb{C})$ is a Banach algebra. Denote by $M_n(\mathbb{R}_o^+)$ the semi-algebra of all matrices belonging to $M_n(\mathbb{C})$ that have non-negative entries. This semi-algebra is locally compact, strict, simple and it has a unit element. The next theorem shows that $M_n(\mathbb{R}_o^+)$ is characterized by these four properties.

3.1 THEOREM. *Let* A *be a locally compact, strict, simple semi-algebra with a unit element. Then there exists a positive integer* n *such that* A *is isomorphic to the semi-algebra* $M_n(\mathbb{R}_o^+)$.

The proof goes in a number of steps. In the sequel A is a locally compact, strict, simple semi-algebra with unit element 1. Since $A \neq (0)$, we know that $1 \neq 0$.

Given non-empty subsets B and D of A, we denote by BD the set of all finite sums

$$b_1d_1 + \dots + b_nd_n$$

with $b_i \in B$ and $d_i \in D$.

3.2 *Let* M *be a non-zero right ideal in* A. *Then* $M^2 \neq (0)$.

PROOF. Suppose $M^2 = (0)$. Then

$$(AM)(AM) \subset AM^2 = (0) \tag{1}$$

Note that AM is a two-sided ideal, and $AM \neq (0)$ since $1 \in A$. But A is simple, therefore AM = A. Then (1) implies that $A^2 = (0)$, contradicting the fact that $0 \neq 1 \in A^2$. Hence $M^2 \neq (0)$.

Since A is locally compact, Lemma 2.1 implies the existence of a minimal closed right ideal in A. By what we have just proved, each minimal closed right ideal in A satisfies the conditions of Theorem 2.5. Hence there exist right minimal idempotents in A (see Proposition 2.3). Let Π denote the set of all right minimal idempotents in A. Then $\Pi \neq \emptyset$.

3.3. $\Pi A = A$.

PROOF. Observe that ΠA is a non-zero right ideal of A. Since A is simple, it suffices to show that ΠA is a left ideal, that is

$$A(\Pi A) \subset \Pi A \tag{2}$$

Take p in Π and a in A. If ap = 0, then $apx = 0 \in \Pi A$ for each x in A. Suppose $ap \neq 0$. Then $(ap)(pA) \neq (0)$, and we can apply Proposition 2.4 to show that (ap)(pA) is a minimal closed right ideal in A. Each minimal closed right ideal in A satisfies the conditions of Theorem 2.5. Hence there exists $q \in \Pi$ such that (see Proposition 2.3)

$$(ap)(pA) = qA.$$

But then ap = qb for some b in A, and thus $apx = q(bx) \in \Pi A$ for each x in A. Hence (2) holds.

3.4. *For* p *and* q *in* Π*, the set* $pAq \neq (0)$.

PROOF. Suppose $pAq = (0)$. Then $q \in (pA)_r$. The set $(pA)_r$ is a two-sided ideal containing a non-zero element, namely q. Therefore $(pA)_r = A$ since A is simple. However, $p \notin (pA)_r$. Contradiction.
Thus $pAq \neq (0)$.

3.5. *Given* p *and* q *in* Π*, there exists a non-zero* u *in* A *such that*

$$pAq = \mathbb{R}_0^+ u.$$

If p = q*, then we may take* u *to be* p.

PROOF. Take $v = qap \neq 0$ in qAp. Then $v(pA) \neq (0)$ and thus, by Proposition 2.4, the set v(pA) is a minimal closed right ideal of A. From $v(pA) \subset qA$ we see that v(pA) = qA, and therefore

$$v(pAq) = qAq.$$

Hence there exists u in pAq such that vu = q. Observe that

$$(pAq)v = pAqap \subset pAp.$$

According to Theorems 2.6 and 2.7, $pAp = \mathbb{R}_o^+ p$. So

$$pAq = (pAq)vu \subset \mathbb{R}_o^+ pu = \mathbb{R}_o^+ u.$$

On the other hand $u \in pAq$, and therefore $\mathbb{R}_o^+ u \subset pAq$. Hence $pAq = \mathbb{R}_o^+ u$.

If p = q, then by Theorems 2.6 and 2.7 we may take u to be p.

3.6. *There exists* $u_1, u_2, \ldots, u_n$ *in* Π *such that*

$$1 = u_1 + u_2 \ldots + u_n, \quad u_i u_j = 0 \quad (i \neq j).$$

PROOF. Because $1 \in A$ and $\Pi A = A$ (see 3.3), there exist e_i in Π and a_i in A (i = 1,2,...,n) such that

$$1 = e_1 a_1 + e_2 a_2 + \ldots + e_n a_n. \tag{3}$$

We may suppose that the representation (3) of 1 is chosen so that n is as small as possible. Postmultiplying (3) by e_1 gives

$$e_1 = e_1 a_1 e_1 + e_2 a_2 e_1 + \ldots + e_n a_n e_1.$$

By 3.5 we have $e_1 a e_1 = \lambda e_1$ with $\lambda \geq 0$. Hence

$$(1 - \lambda)e_1 = e_2 a_2 e_1 + \ldots + e_n a_n e_1. \tag{4}$$

Formula (4) implies that $\lambda \geq 1$, for, if $\lambda < 1$, then we can substitute for e_1 in (3) and we obtain a representation of the unit element of the form

$$1 = e_2 b_2 + \ldots + e_n b_n$$

say, which contradicts the definition of n. Therefore $\lambda \geq 1$. On the other hand, because of the strictness of A, formula (4) shows that $\lambda > 1$ is impossible. Thus $\lambda = 1$ and hence

$$e_1 a e_1 = e_1, \; e_2 a_2 e_1 + \ldots + e_n a_n e_1 = 0.$$

Again using the strictness of A, we see that the second part of the last formula implies that $e_i a_i e_1 = 0$ for i = 2,...,n. By replacing e_1 by e_j in the preceding arguments, we obtain

$$e_j a_j e_j = e_j, \quad e_i a_i e_j = 0 \quad (i \neq j).$$

Let $u_i = e_i a_i$ (i = 1,...,n). Then we have proved that

$$1 = u_1 + \dots + u_n, \quad u_i^2 = u_i, \quad u_i u_j = 0 \quad (i \neq j).$$

Since $0 \neq u_i = u_i^2 = u_i(e_i a_i)$, we have $u_i(e_i A) \neq (0)$, and so we can apply Proposition 2.4 to show that $u_i(e_i A)$ is a minimal closed right ideal. Clearly, $u_i \in u_i(e_i A)$. But then, by Proposition 2.3, $u_i \in \Pi$.

3.7. *For each* i *and* j *we have*

$$u_i A u_j = \mathbb{R}_0^+ e_{ij}, \tag{5}$$

where e_{ij} *is some non-zero element of* $u_i A u_j$. *The elements* e_{ij} *can be chosen so that*

$$e_{ii} = u_i \ (i = 1,\dots,n),\ e_{ij}e_{jk} = e_{ik} \ (i,j,k = 1,\dots,n),$$

$$e_{ij}e_{kh} = 0 \quad (j \neq k).$$

PROOF. Since $u_i \in \Pi$, we know from 3.5 that there exist elements e_{ij} satisfying (5). Moreover, again according to 3.5, we can choose e_{ii} to be u_i for $i = 1,\dots,n$. Observe that the condition

$$e_{ij}e_{kh} = 0 \quad (j \neq k)$$

is automatically satisfied since $u_j u_k = 0$ for $j \neq k$. It remains to choose the e_{ij} so that $e_{ij}e_{jk} = e_{ik}$.

For the elements e_{1j} with $j \neq 1$, we take arbitrary non-zero elements of $u_1 A u_j$. Then (5) holds for these elements. Next, let $j \neq 1$ and k be arbitrary. The left annihilator $(Au_k)_\ell$ is a two-sided ideal not containing u_k. Since A is simple, this implies that $(Au_k)_\ell = (0)$. Therefore

$$(0) \neq e_{1j}Au_k = e_{1j}u_j A u_k \subset u_1 A u_k$$

and hence

$$e_{1j}u_j A u_k = u_1 A u_k.$$

This shows that the equation

$$e_{1j}X = e_{1k} \tag{6}$$

has one (and no more than one) solution in $u_j A u_k$, e_{jk} say. We have now completed the choice of the e_{jk}. Observe that for $j = k \geq 2$ the solution of (6) is u_j. Further $e_{jk} \neq 0$ and belongs to $u_j A u_k$. Thus (5) holds.

According to (6), we have $e_{1j}e_{jk} = e_{1k}$ for $j \neq 1$. Since $e_{11}e_{1k} = e_{1k}$, we obtain

$$e_{1j}e_{jk} = e_{1k} \quad (j,k = 1,\dots,n).$$

Take arbitrary i, j and k. From

$$e_{ij}e_{jk} \in u_i A u_k = \mathbb{R}_o^+ e_{ik},$$

we conclude that $e_{ij}e_{jk} = \lambda e_{ik}$. But then

$$e_{1k} = e_{1i}(e_{ij}e_{jk}) = \lambda e_{1i}e_{ik} = \lambda e_{1k}.$$

This implies that $\lambda = 1$, and hence $e_{ij}e_{jk} = e_{ik}$ for all i, j and k.

We are now able to prove Theorem 3.1. Take x in A. Then

$$x = 1.x.1 = \sum_{i,j=1}^{n} u_i x u_j = \sum_{i,j=1}^{n} \lambda_{ij}e_{ij}.$$

The map $x \to (\lambda_{ij})_{i,j=1}^{n}$ is the desired isomorphism.

NOTES ON CHAPTER II

1. In [4] Bonsall and Tomiuk have proved that the monothetic semi-algebra generated by a compact linear operator T which has the additional property that

$$0 < r(T) \in \sigma(T),$$

is locally compact. This result is the first version of Theorem 1.1. A second version of the theorem appears in [19], where it was shown that the condition that T is compact can be replaced by the requirement that the peripheral spectrum of T consists of poles of T.

2. It is easy to construct an example of a compact operator T with $r(T) = 0$ which generates a semi-algebra which is not locally compact. Let ℓ_1 be the Banach space of absolutely summable sequences $x = \{x_n\}$ of complex numbers, and let T on ℓ_1 be defined by

$$(Tx)_n = \frac{x_{n+1}}{n+1} \quad (n = 1,2,\ldots).$$

Then T is compact and quasi-nilpotent (i.e., $r(T) = 0$), but $A(T)$ is not locally compact (see [19] for details).

3. Each nilpotent operator generates a semi-algebra which is locally compact. Whether a quasi-nilpotent, non-nilpotent linear operator can generate a locally compact semi-algebra is an open problem.

4. The first example of an infinite dimensional locally compact semi-algebra is given by Bonsall in [2]. It goes as follows. Let Δ be the set of real numbers consisting of the closed unit interval $[0,1]$ together with the real number 2, and let Δ be given the relative topology induced from the usual topology on the real line. Denote by A' the class of real-valued functions defined on the closed interval $[0,2]$ that are continuous, non-negative, increasing and convex there. Finally, let A denote the class of functions on Δ that are the restrictions to Δ of functions in A'. Then A is a locally compact semi-algebra not contained in a finite dimensional algebra (see [2] for details).

5. Theorem 1.2 is related to work of L. Elsner. In [11] Elsner proved the following theorem. Let T be a compact linear operator on a complex Banach space E. Suppose that $r(T) > 0$ and that $r(T)$ is a pole of T of maximal order in the peripheral spectrum of T. Then there exists a closed and total cone (see section IV.2 for the definitions of these notions) K in E such that $TK \subset K$. The fact that K is a closed and total T-invariant cone implies that $A(T)$ is strict (see Chapter IV for more results of this type). Hence for operators T of this kind the first part of Theorem 1.2 follows from Elsner's theorem (cf. Theorem 8 in [17]). Whether or not Elsner's theorem holds for operators T satisfying the conditions of Theorem 1.2 is an unsolved problem.

CHAPTER III

SEMI-SIMPLE LOCALLY COMPACT MONOTHETIC SEMI-ALGEBRAS

The aim of this chapter is to obtain a converse of Theorem II.1.1 which gives sufficient conditions for an element in a Banach algebra to generate a locally compact semi-algebra. We do not completely achieve this, but we make considerable progress culminating in Theorem 3.4 which characterizes semi-simple locally compact semi-algebras in terms of the spectrum of a generator. For a general semi-algebra semi-simplicity is a condition on the closed two-sided ideals which, in the commutative locally compact case, is equivalent to the requirement that the spectral radius of every non-zero element be non-zero (Proposition 1.3). Theorem 2.1 characterizes strict semi-simple locally compact monothetic semi-algebras by the conditions that the spectral radius of the generator be non-zero and in the spectrum, and that its peripheral spectrum be a set of simple poles. Proposition 2.3 links strict semi-simple locally compact monothetic semi-algebras to the compact monothetic semi-group with the same generator. This opens the way for using Theorem I.2.3 (which characterizes compact monothetic semigroups) to characterize strict semi-simple locally compact monothetic semi-algebras. This technique was originally used by the authors in [19].

In section 3 the strictness condition is omitted, and the extra complications which this entails are finally disposed of in Theorem 3.4. The result may be loosely stated as follows. Necessary and sufficient conditions that a non-zero element of a Banach algebra should generate a semi-simple locally compact semi-algebra are that, apart from a finite set of simple poles lying off the positive real axis, the peripheral spectrum of the generator consists of a set of simple poles and includes the spectral radius. Theorem 3.4 is due to the authors (see Theorem B in [20]).

The second main result in this chapter is a representation theorem (Theorem 2.7) for strict semi-simple locally compact monothetic semi-algebras which generalizes Bonsall and Tomiuk's Theorem 11 in [4]. It states that each element in such a semi-algebra is an infinite sum of positive multiples of powers of the generator plus a finite linear combination of elements from the kernel of the associated semigroup. One also finds that the idempotents in such a semi-algebra form a finite set (Corollary 2.10). In section 3 we show that the set of idempotents remains finite if the strictness condition is omitted (Corollary 3.5). These last results are new.

1. DEFINITIONS AND ELEMENTARY PROPERTIES

Throughout this chapter B will be a complex Banach algebra with unit e. A semi-algebra A (in B) is said to be *semi-simple* if the zero ideal is the only closed two-sided ideal J in A with $J^2 = (0)$.

1.1 LEMMA. *Let* A *be a semi-simple semi-algebra, and let* I *be an ideal (left, right, or two-sided) of* A *such that* $I^n = (0)$ *for some positive integer* n. *Then* $I = (0)$.

PROOF. Let I be a left ideal of A with $I^2 = (0)$. Take J to be the closure in A of the set IA. Then J is a closed two-sided ideal in A. Since

$$(IA)(IA) \subset I^2A = (0),$$

we have $J^2 = (0)$. Hence, by semi-simplicity, $J = (0)$ and thus $IA = (0)$. This shows that I is a subset of the left annihilator A_ℓ of A. Since A_ℓ is a closed two-sided ideal in A with $A_\ell^2 \subset A_\ell A = (0)$, we have $A_\ell = (0)$. Hence $I = (0)$. This proves the lemma for the case that $n = 2$.

Now let I be a left ideal with $I^n = (0)$ for an arbitrary positive integer n. Let k be the least positive integer with $I^k = (0)$. If $k > 1$, we have

$$(I^{k-1})^2 \subset I^{k-1}I = (0).$$

Since I^{k-1} is a left ideal, the first part of the proof shows that $I^{k-1} = (0)$, which is absurd. Hence $k = 1$, that is, $I = (0)$. A similar argument applies to right ideals.

1.2 PROPOSITION. *Let* A *be a commutative semi-algebra. Then* A *is semi-simple if and only if* $a^2 \neq 0$ *for each non-zero* a *in* A.

PROOF. Suppose that A is semi-simple. Take a in A with $a^2 = 0$, and consider the set

$$I = \{ba + \alpha a : b \in A, \alpha \geq 0\}.$$

Because of the commutativity of A, I is an ideal in A with $I^2 = (0)$. Thus $I = (0)$. Since $a \in I$, this implies $a = 0$. The converse is trivial.

The next proposition gives a necessary and sufficient condition for semi-simplicity in terms of the spectral radii of the non-zero elements in the semi-algebra. It also links the semi-simplicity of semi-algebras to the analogous notion for commutative Banach algebras. Recall that r(a) denotes the spectral radius of the element a in B.

1.3 PROPOSITION. *Let* A *be a locally compact commutative semi-algebra. Then* A *is semi-simple if and only if*

$$r(a) > 0 \quad (0 \neq a \in A).$$

PROOF. Suppose that A is semi-simple. Consider the set

$$I = \{x \in A : r(x) = 0\}.$$

From the properties of the spectral radius on commutative subsets of B (Proposition P.2) we see that I is a closed two-sided ideal in A. Suppose that $I \neq (0)$. Then, by Lemma II.2.1, I contains a minimal closed ideal, M say. Since A is semi-simple, M satisfies the conditions of Theorem II.2.5. Hence M contains a non-zero idempotent. But any non-zero idempotent has spectral radius one. Contradiction. Therefore $I = (0)$, and thus $r(a) > 0$ for any non-zero a in A. The converse is trivial.

Observe that a simple semi-algebra (see section II.3) is not necessarily semi-simple. E.g., take t to be a nilpotent element in B with order of nilpotence equal to 2, and let

$$A = \{\alpha t: \alpha \geq 0\}.$$

Then A is a simple semi-algebra, but A is not semi-simple.

2. SEMI-SIMPLICITY AND STRICTNESS

In this section we give a complete characterization of strict, semi-simple locally compact monothetic semi-algebras in terms of the spectrum of a generating element. The theorem we shall prove reads as follows.

2.1 THEOREM. *Let* t *be an element in* B. *Then* A(t) *is locally compact, semi-simple and strict if and only if*

(i) $0 < r(t) \in \sigma(t)$,

(ii) $\mathrm{Per}\sigma(t)$ *is a finite set of simple poles of* t.

The proof of the theorem is based on two propositions. The first is a corollary to Proposition 1.3.

2.2 PROPOSITION. *Let* A *be a commutative, semi-simple locally compact semi-algebra. Then there is a positive constant* C *such that*

$$\|a\| \leq Cr(a) \quad (a \in A).$$

PROOF. Let $A_1 = \{x \in A: \|x\| = 1\}$. Then A_1 is a compact set. Since A is commutative, the function

$$x \to r(x)$$

is continuous on A (by Proposition P.2) and hence on A_1. From Proposition 1.3 we know that $r(x) > 0$ for each $x \in A_1$. So there exists $\varepsilon > 0$ such that $r(x) \geq \varepsilon > 0$ for each x in A_1. But then

$$r(x) \geq \varepsilon\|x\| \quad (x \in A).$$

Take $C = \varepsilon^{-1}$, and the proof is complete.

Let t be an element in B with $r(t) = 1$. Suppose that A(t) is locally compact and semi-simple. Since A(t) is also commutative, we get from the preceding result the existence of a positive constant C such that

$$\|t^n\| \leq Cr(t^n) \quad (n = 1,2,\ldots).$$

But $r(t^n) = r(t)^n$ for each n. Hence

$$\|t^n\| \leq C \quad (n = 1,2,\ldots).$$

Since A(t) is locally compact, this implies that the set

$$S(t) = \mathrm{cl}\{t,t^2,\ldots\}$$

is a compact subset of B. The next proposition gives further information about the relationship between the locally compact semi-algebra A(t) and the compact semigroup S(t). The proof of this proposition leans heavily on the results of the sections I.2, II.1 and II.2.

2.3 PROPOSITION. *Let* t *be an element in* B *with* $r(t) = 1$. *Then the following statements are equivalent.*

(i) A(t) *is locally compact, semi-simple and strict.*

(ii) S(t) *is compact and* $1 \in \sigma(t)$.

PROOF. (i) $\Rightarrow$ (ii). We proved already that S(t) is compact. So we can apply Theorem I.2.3 to show that

$$\mathrm{Per}\sigma(t) = \{\lambda \in \sigma(t)\colon |\lambda| = r(t) = 1\}$$

is a set of simple poles of t. In particular $\mathrm{Per}\sigma(t)$ is a spectral set of t. From Theorem I.2.4 we know that the spectral idempotent q associated with $\mathrm{Per}\sigma(t)$ is an element of S(t), and thus $q \in A(t)$. Let

$$I = \{xq\colon x \in A(t)\}.$$

Then I is a closed ideal in A(t). Since $\mathrm{Per}\sigma(t) \neq \emptyset$, standard spectral theory implies that $tq \neq 0$ (cf., Proposition P.5). Hence $I \neq (0)$.

Let M be a minimal closed ideal contained in I. Such an ideal exists because of Lemma II.2.1 and the fact that A(t) is locally compact and commutative. Since A(t) is semi-simple, $M^2 \neq (0)$, and we can apply Theorem II.2.5 to show that M contains a non-zero idempotent, p say. Observe that p is a minimal idempotent in the commutative semi-algebra A(t). Hence by Theorem II.2.6, pA(t) is a closed division semi-algebra with unit element p. But A(t) is strict, hence it follows from Theorem II.2.7 that

$$pA(t) = \mathbb{R}_o^+ p.$$

In particular, $tp = \alpha p$ for some $\alpha \geq 0$.

Suppose $\alpha = 0$. Then $t^n p = 0$ for $n = 1,2,\ldots$, and hence $sp = 0$ for each s in A(t). In particular, $p^2 = 0$. But this contradicts the fact that $p^2 = p \neq 0$. Thus $\alpha > 0$. Since $p \in I$, $pq = q$, and hence

$$(\alpha - tq)p = 0.$$

This shows that $0 < \alpha \in \sigma(tq)$. From Proposition P.5 we know that the non-zero part of $\sigma(tq)$ coincides with $\text{Per}\sigma(t)$. Thus $0 < \alpha \in \text{Per}\sigma(t)$, that is, $\alpha = 1 \in \sigma(t)$.

(ii) ⟹ (i). From our hypothesis and Theorem I.2.3 it follows that $\text{Per}\sigma(t)$ is a set of simple poles of t and $1 \in \text{Per}\sigma(t)$. Hence t satisfies the conditions of Theorems II.1.1 and II.1.2. So $A(t)$ is locally compact and strict. It remains to show that $A(t)$ is semi-simple.

Let p be a polynomial with non-negative coefficients. Since $S(t)$ is compact, there exists a positive constant C such that

$$\|t^n\| \leq C \quad (n = 1,2,\ldots).$$

This implies that

$$\|p(t)\| \leq Cp(1). \tag{1}$$

Using the spectral mapping theorem (Proposition P.3), we see that $r(t) = 1 \in \sigma(t)$ implies that

$$r(p(t)) = p(1). \tag{2}$$

Since the spectral radius $r(s)$ is a continuous function of s on the commutative set $A(t)$ (Proposition P.2), formulas (1) and (2) together show that

$$\|s\| \leq Cr(s) \quad (s \in A(t)).$$

But this implies that $r(s) > 0$ for each non-zero $s \in A(t)$. Hence, by Proposition 1.3, $A(t)$ is semi-simple.

A normalization is all that is required to complete the proof of Theorem 2.1. Suppose that $A(t)$ is locally compact, semi-simple and strict. Local compactness of $A(t)$ implies that $A(t) \neq (0)$ and thus $t \neq 0$. But then we can apply Proposition 1.3 to show that $r(t) > 0$. Let

$$s = \frac{1}{r(t)}\, t. \tag{3}$$

Then $A(s) = A(t)$ and s and t have the same spectral properties. Hence it suffices to prove (i) and (ii) in Theorem 2.1 for s instead of t. Since $r(s) = 1$, we can apply Proposition 2.3 to show that

$$1 = r(s) \in \sigma(s) \tag{4}$$

and that $S(s)$ is compact. According to Theorem I.2.3, the last fact implies that Perσ(s) is a set of simple poles of s. This proves the necessity of the conditions (i) and (ii) of Theorem 2.1.

To prove the sufficiency of the conditions, we take s as in formula (3). Then (4) holds, and Perσ(s) is a set of simple poles of s. So we can apply Theorem I.2.3 to show that S(s) is compact. But then, using Proposition 2.3, we see that A(t) = A(s) is locally compact, semi-simple and strict. This completes the proof of Theorem 2.1.

Proposition 1.2 makes it easy to construct examples of semi-simple semi-algebras. E.g., let t be a non-zero element in the complex Banach algebra $\mathbb{C}^n$,

$$t = (t_1, t_2, \dots, t_n),$$

say. Since the square of a non-zero element in $\mathbb{C}^n$ is non-zero, Proposition 1.2 shows that A(t) is semi-simple. The fact that $\mathbb{C}^n$ is finite dimensional implies that A(t) is locally compact, and hence Theorem 2.1 gives a necessary and sufficient condition for the strictness of A(t). In this particular case the spectrum of t is the set $\{t_1, \dots, t_n\}$, each t_i is a simple pole of t and

$$r(t) = \max_i |t_i| > 0.$$

Hence we know from Theorem 2.1 that in this case the following two conditions are equivalent.

(i) A(t) is strict.

(ii) $\max_i |t_i| \in \{t_1, \dots, t_n\}$.

A less trivial application of Theorem 2.1 will be given in Chapter IV, where we shall show that an irreducible positive operator generates a semi-simple strict semi-algebra.

We proceed with a further investigation of the inclusion $S(t) \subset A(t)$.

2.4 PROPOSITION. *Let* t *be an element in* B *such that* A(t) *is locally compact, semi-simple and strict. Then the set*

$$\Sigma(t) = \{x \in A(t) : r(x) = 1\}$$

is a compact, convex subset of A(t).

PROOF. From Theorem 2.1 we know that $0 < r(t) \in \sigma(t)$ and that Perσ(t) is a set of simple poles of t. Without loss of generality we may suppose that $r(t) = 1$. Let e_o be the spectral idempotent corresponding to the spectral set $\{1\}$. Then $te_o = e_o$, and hence

$$p(t)e_o = p(1)e_o$$

for any polynomial p. Let p be a polynomial with non-negative coefficients. Then we know (see formula (2)) that $r(p(t)) = p(1)$, and thus

$$p(t)e_o = r(p(t))e_o. \tag{5}$$

Since the spectral radius $r(s)$ is a continuous function of s on the commutative set $A(t)$ (Proposition P.2), we obtain from formula (5) that

$$se_o = r(s)e_o \quad (s \in A(t)). \tag{6}$$

Now take x_1 and x_2 in $\Sigma(t)$, and let $0 < \lambda < 1$. Then

$$\begin{aligned}\{\lambda x_1 + (1 - \lambda)x_2\}e_o &= \lambda r(x_1)e_o + (1 - \lambda)r(x_2)e_o\\ &= \lambda e_o + (1 - \lambda)e_o\\ &= e_o.\end{aligned}$$

On the other hand

$$\{\lambda x_1 + (1 - \lambda)x_2\}e_o = r(\lambda x_1 + (1 - \lambda)x_2)e_o.$$

Thus from (6) we obtain $r(\lambda x_1 + (1 - \lambda)x_2) = 1$, and hence $\lambda x_1 + (1 - \lambda)x_2 \in \Sigma(t)$. This shows that $\Sigma(t)$ is convex.

Next observe that the continuity of the spectral radius implies that $\Sigma(t)$ is closed. Further, by Proposition 2.2, $\Sigma(t)$ is a bounded subset of $A(t)$. Since $A(t)$ is locally compact, it follows that $\Sigma(t)$ is compact.

Let t be as in the previous proposition, and suppose that $r(t) = 1$. Then

$$\Lambda(t) = \{t, t^2, \ldots\} \subset \Sigma(t).$$

We shall prove that $\Sigma(t)$ is the closed convex hull of $\Lambda(t)$.

2.5 PROPOSITION. *Let* t *be an element in* B *such that* $A(t)$ *is locally compact, semi-simple and strict. Suppose that* $r(t) = 1$. *Then* $\Sigma(t)$ *is the closed convex hull of* $\Lambda(t)$.

PROOF. Take x in $\Sigma(t)$. Then $x \in A(t)$, and hence x is the limit of a sequence $\{p_n(t)\}$ of polynomials in t with non-negative coefficients and first coefficient zero. From the continuity of the spectral radius we know that

$$1 = r(x) = \lim r(p_n(t)).$$

Hence, without loss of generality, we may suppose that $r(p_n(t)) > 0$. Under the conditions of the proposition we have (see formula (2))

$$r(p_n(t)) = p_n(1).$$

Let

$$q_n(t) = \frac{1}{p_n(1)} p_n(t).$$

Then $q_n(t)$ is a convex combination of powers of t. Further

$$x = \lim q_n(t).$$

Thus x is in the closed convex hull Γ of the set $\Lambda(t)$. Clearly, by Proposition 2.4, $\Gamma \subset \Sigma(t)$. So we have proved that $\Sigma(t) = \Gamma$.

An immediate consequence of the preceding proposition is that $\Sigma(t)$ is the closed convex hull of

$$S(t) = \mathrm{cl}\{t, t^2, \ldots\}.$$

In the next theorem we shall give more information about the closed convex hull of S(t), but first we prove a lemma about the closed convex hull of K. Here K denotes the kernel of the compact monothetic semi-group S(t) (see section I.1). Recall that

$$S(t) = K \cup \Lambda(t).$$

2.6 LEMMA. *Let* t *be an element in* B *such that* A(t) *is locally compact, semi-simple and strict. Suppose that* $r(t) = 1$. *Then the closed convex hull of* K *is the set of all elements of the form*

$$\alpha_1 g_1 + \alpha_2 g_2 + \ldots + \alpha_{2m+1} g_{2m+1}$$

with $\alpha_j \geq 0$, $g_j \in K$ $(j = 1, \ldots, 2m+1)$,

$$\sum_{j=1}^{2m+1} \alpha_j = 1,$$

and where m *is the number of points in* $\mathrm{Per}\sigma(t)$.

PROOF. From Theorem 2.1 we know that $\mathrm{Per}\sigma(t)$ is a finite set of simple poles of t. If

$$\operatorname{Per}\sigma(t) = \{\lambda_1,\lambda_2,\ldots,\lambda_m\},$$

let p_j be the spectral idempotent corresponding to the spectral set $\{\lambda_j\}$. Then

$$q = p_1 + p_2 + \ldots + p_m$$

is the kernel idempotent of S(t) (see Theorem I.2.4) and

$$tq = \lambda_1 p_1 + \ldots + \lambda_m p_m.$$

This implies that K is contained in a linear subspace of B of dimension 2m over $\mathbb{R}$. Therefore its convex hull co(K) is the set of elements of the stated form (see [26], section 1). Since K is compact and the map

$$(g_1,\ldots,g_{2m+1}, \alpha_1,\ldots,\alpha_{2m+1}) \to \sum_{j=1}^{2m+1} \alpha_j g_j$$

is continuous, co(K) is compact and therefore closed. This completes the proof of the lemma.

2.7 THEOREM. *Let* t *be an element in* B *such that* A(t) *is locally compact, semi-simple and strict, and suppose that* r(t) = 1. *Let* m *be the number of points in* Perσ(t). *Then* Σ(t) *is the set of all elements of the form*

$$\sum_{k=1}^{\infty} \alpha_k t^k + \sum_{j=1}^{2m+1} \beta_j g_j$$

with $\alpha_k \geq 0$ $(k = 1,2,\ldots)$, $\beta_j \geq 0$, $g_j \in K$ $(j = 1,\ldots,2m+1)$ *and*

$$\sum_{k=1}^{\infty} \alpha_k + \sum_{j=1}^{2m+1} \beta_j = 1. \tag{7}$$

PROOF. It is clear that all elements of the stated form are in Σ(t). To prove the converse, take x in Σ(t). By Proposition 2.5, x is the limit of a sequence $\{x_n\}$ in the convex hull of Λ(t). That is, for each n,

$$x_n = \sum_{k=1}^{\infty} \alpha(n,k)t^k$$

with $\alpha(n,k) \geq 0$ $(n,k = 1,2,\ldots)$, $\alpha(n,k) = 0$ for k sufficiently large and

$$\sum_{k=1}^{\infty} \alpha(n,k) = 1.$$

Using the diagonal process, and then throwing away all but a suitable subsequence, we may suppose that, for each fixed k, the sequence $\{\alpha(n,k)\}$ converges to α_k, say. Then

$$\sum_{k=1}^{\infty} \alpha_k \leq 1.$$

Let

$$a = \sum_{k=1}^{\infty} \alpha_k t^k.$$

Since S(t) is compact, t has equibounded iterates, and thus a is a well-defined element in A(t).

Given a positive integer r, let

$$x(r,n) = \sum_{k=1}^{r} \alpha(n,k)t^k, \quad a(r) = \sum_{k=1}^{r} \alpha_k t^k.$$

Then

$$\lim_n x(r,n) = a(r) \quad (r = 1,2,\ldots). \tag{8}$$

From our hypothesis it follows that Per$\sigma(t)$ is a spectral set of t (cf., Theorem 2.1). Let p be the associated spectral idempotent. Then $r(t - tp) < 1$, and hence

$$t^k(e - p) = (t - tp)^k \to 0 \quad (k \to +\infty).$$

Take $\varepsilon > 0$. There exists a positive integer k_0 such that $\|t^k(e - p)\| < \varepsilon$ for $k \geq k_0$. Clearly, this implies that

$$\left\| \sum_{k=k_0+1}^{\infty} \alpha(n,k)t^k(e - p) - \sum_{k=k_0+1}^{\infty} \alpha_k t^k(e - p)\right\| < 2\varepsilon.$$

Further, we see from formula (8) that

$$\lim_n x(k_0,n)(e - p) = a(k_0)(e - p).$$

Combining these two results we obtain

$$x(e - p) = \lim_n x_n(e - p) = a(e - p). \tag{9}$$

Observe that

$$x_n - x(r,n) \in A(t) \quad (n,r = 1,2,\ldots).$$

Since A(t) is closed, we see from $x = \lim x_n$ and formula (8) that

$$x - a(r) \in A(t) \quad (r = 1,2,\ldots).$$

But then, using the fact that $a = \lim a(r)$, we obtain

$$b = x - a \in A(t).$$

Let Γ be the closed convex hull of the kernel K of S(t). From Theorem

I.2.4 we know that p belongs to K. Therefore, $t^n p \in K$ for $n = 1,2,\ldots$. Hence for any polynomial q with non-negative coefficients and first coefficient zero, we have

$$q(t)p \in \{\alpha u:\ \alpha \geq 0,\ u \in \Gamma\}.$$

But this implies that

$$A(t)p \subset \{\alpha u:\ \alpha \geq 0,\ u \in \Gamma\}.$$

Formula (9) and the definition of b imply that $b(e - p) = 0$, that is $b = bp$. Hence $b \in A(t)p$ and therefore

$$b \in \{\alpha u:\ \alpha \geq 0,\ u \in \Gamma\}.$$

But then we can apply Lemma 2.6, to show that b is of the form

$$b = \sum_{j=1}^{2m+1} \beta_j g_j$$

with $\beta_j \geq 0$ and $g_j \in K$ $(j = 1,\ldots,2m+1)$. It remains to show that (7) holds.

From our hypothesis it follows that $\{1\}$ is a spectral set of t (cf. Theorem 2.1). Let e_o be the corresponding spectral idempotent. Then we know (see formula (6)) that

$$se_o = r(s)e_o \qquad (s \in A(t)).$$

For each element s in S(t), $r(s) = 1$. This implies that

$$xe_o = \left(\sum_{k=1}^{\infty} \alpha_k + \sum_{j=1}^{2m+1} \beta_j\right)e_o,$$

and thus, since $r(x) = 1$, (7) holds.

Let t be as in the preceding theorem. If s is any non-zero element in A(t), then, by semi-simplicity, $r(s) > 0$, and thus

$$s_1 = \frac{1}{r(s)}\, s \in \Sigma(t).$$

Hence s is of the form αs_1 with $s_1 \in \Sigma(t)$ and $\alpha \geq 0$. Clearly, this representation is unique. This, together with Proposition 2.4, shows that $\Sigma(t)$ is a compact convex base for A(t). Further, using the result of Theorem 2.7, it implies that A(t) is the set of all elements s of the form

$$s = \sum_{k=1}^{\infty} \alpha_k t^k + \sum_{j=1}^{2m+1} \beta_j g_j$$

with $\alpha_k \geq 0$ $(k = 1,2,\ldots)$, $\beta_j \geq 0$, $g_j \in K$ $(j = 1,\ldots,2m+1)$ and

$$r(s) = \sum_{k=1}^{\infty} \alpha_k + \sum_{j=1}^{2m+1} \beta_j .$$

Here m is as in Theorem 2.7. Observe that this representation theorem shows that the spectral radius is additive on A(t). This result is also visible from formula (6).

The spectral idempotent e_o corresponding to the spectral set {1} is an element of A(t). In fact (see Proposition I.2.1)

$$e_o = \lim_n \frac{1}{n} (t + \dots + t^n).$$

Since (see formula (6))

$$se_o = r(s)e_o \qquad (s \in A(t)),$$

the set

$$M = \mathbb{R}_o^+ e_o$$

is a minimal closed ideal and e_o is the corresponding minimal idempotent. It is interesting to observe that A(t) does not contain any other minimal closed ideal. Indeed, let N be a minimal closed ideal in A(t) different from M. The intersection $N \cap M$ is a closed ideal properly contained in M. Hence $N \cap M = (0)$. Let s be a non-zero element in N. Then $r(s) > 0$ and hence

$$se_o = r(s)e_o \neq 0,$$

contradicting the fact that $NM \subset N \cap M = (0)$. Thus M is the only minimal closed ideal in A(t). Since M contains precisely one non-zero idempotent, namely e_o, e_o is the only minimal idempotent in A(t).

Observe that the spectral idempotent p corresponding to the spectral set $\mathrm{Per}\sigma(t)$ belongs to A(t). In fact p is the unit of the kernel K of S(t). In general, $p \neq e_o$. It can happen that A(t) contains more than two idempotents.

2.8 EXAMPLE. Take $B = \mathbb{C}^4$, and let

$$t = \{1, i, -1, -i\},$$

where $i^2 = -1$. Then A(t) is locally compact, semi-simple and strict (see page 50). Further, $r(t) = 1$. So t satisfies the conditions of Theorem 2.7. Observe that

$$\tfrac{1}{4}(t + t^2 + t^3 + t^4) = \{1,0,0,0\}, \; t^4 = \{1,1,1,1\}$$

and

$$\tfrac{1}{2}(t^2 + t^4) = \{1,0,1,0\}.$$

Thus A(t) contains the three idempotents $e_o = \{1,0,0,0\}$, $p = \{1,1,1,1\}$ and $q = \{1,0,1,0\}$. Note that

$$e_oq = e_o, \quad pq = q.$$

The next theorem shows that the last formula holds in general.

2.9 PROPOSITION. *Let* t *be an element in* B *such that* A(t) *is locally compact, semi-simple and strict, and suppose that* $r(t) = 1$. *Let* e_o *be the minimal idempotent in* A(t), *and let* p *be the unit of the kernel* K *of* S(t). *Then* e_o *is the least and* p *is the greatest idempotent in* A(t) *in the sense that*

$$e_oq = e_o, \quad pq = q$$

for every non-zero idempotent q *belonging to* A(t).

PROOF. Let q be a non-zero idempotent in A(t). Then $r(q) = 1$, and hence

$$e_oq = qe_o = r(q)e_o = e_o.$$

Also, by Theorem 2.7,

$$q = \sum_{k=1}^{\infty} \alpha_k t^k + \sum_{j=1}^{2m+1} \beta_j g_j$$

with $\alpha_k \geq 0$ $(k = 1,2,\ldots)$, $\beta_j \geq 0$, $g_j \in K$ $(j = 1,\ldots,2m+1)$ and (7) holds. For each g in K, we have $gp = g$, that is, $g(e - p) = 0$. Thus

$$q(e - p) = \sum_{k=1}^{\infty} \alpha_k t^k(e - p).$$

Put $s = t(e - p)$. Since the spectral radius is an additive continuous function on A(t),

$$r(q(e - p)) = r\Big(\sum_{k=1}^{\infty} \alpha_k s^k\Big) = \sum_{k=1}^{\infty} \alpha_k r(s)^k.$$

Now $r(s) < 1$, and hence it follows from (7) that

$$r(q(e - p)) \leq r(s) < 1.$$

But $q(e - p)$ is an idempotent, therefore $r(q(e - p)) = 0$ and hence $q = qp$.

2.10 COROLLARY. *The number of idempotents in a strict, semi-simple locally compact monothetic semi-algebra is finite.*

PROOF. Let t be as in Proposition 2.9. We have to prove that the number of idempotents in A(t) is finite. Let q be a non-zero idempotent in A(t). Then (by Proposition P.6) q is a spectral idempotent corresponding to a non-empty spectral set σ, say. Since

$$e_o q = e_o, \quad pq = q,$$

it follows from spectral theory that

$$1 \in \sigma \subset \mathrm{Per}\sigma(t).$$

Since Perσ(t) is a finite set, the number of different subsets of Perσ(t) is finite. Therefore, the number of non-zero idempotents in A(t) is finite.

Let t be as in Proposition 2.9, and let m be the number of points in the peripheral spectrum of t. The proof of Corollary 2.10 shows that the number of non-zero idempotents in A(t) is at most 2^{m-1}. However this number may be strictly less than 2^{m-1}. E.g., in Example 2.9, m = 4 and a careful examination of this example shows that in this case A(t) contains precisely 3 idempotents which is strictly less than 2^3.

In the next section we shall show that the strictness condition in Corollary 2.10 is superfluous (see Proposition 3.5).

3. THE NON-STRICT CASE

In this section we give a complete characterization of semi-simple locally compact monothetic semi-algebras in terms of the spectrum of a generating element (see Theorem 3.4). The proof of this theorem is based on Theorem 2.1 of the preceding section and a decomposition theorem for commutative, semi-simple locally compact semi-algebras. Before we state this decomposition theorem, we introduce the following notation. Let A_1 and A_2 be semi-algebras in B. We write

$$A = A_1 \oplus A_2$$

whenever

$$A = \{a_1 + a_2 : a_i \in A_i \ (i = 1,2)\}$$

and each a in A is uniquely representable in the form $a = a_1 + a_2$ with

$a_i \in A_i$ $(i = 1,2)$.

3.1 THEOREM. *Let* A *be a non-zero commutative semi-algebra. Then* A *is locally compact and semi-simple if and only if there exists an idempotent* p *in* A *such that*

(i) $A = Ap \oplus \{p\}_r$;

(ii) *either* $Ap = (0)$ *or* Ap *is locally compact, semi-simple and* $Ap = -Ap$;

(iii) *either* $\{p\}_r = (0)$ *or* $\{p\}_r$ *is locally compact, semi-simple and strict.*

PROOF. Suppose that A is locally compact and semi-simple. If A is strict, then we take $p = 0$, and in that case the conditions (i), (ii) and (iii) are trivially satisfied. Therefore suppose that

$$J = A \cap (-A) \neq (0).$$

Observe that, because of local compactness, J is a finite dimensional real linear space. Further J is a closed ideal in A.

Let M be a minimal closed ideal in A. Since A is semi-simple, M satisfies the conditions of Theorem II.2.5. Hence M contains a non-zero idempotent, e_M say. Since

$$M = e_M A = e_M A e_M$$

is a division semi-algebra (Theorem II.2.6), e_M is the only non-zero idempotent in M. Hence e_M is uniquely determined. If M_1 and M_2 are two distinct minimal closed ideals in A, then

$$M_1 M_2 \subset M_1 \cap M_2 = (0),$$

and hence it follows that

$$e_{M_1} e_{M_2} = 0 \qquad (1)$$

The last formula shows that the set of all e_M, where M is any minimal closed ideal in A, is a linearly independent set.

Let V be the set of all minimal closed ideals of A which are contained in J. From Lemma II.2.1 we know that V is not empty. Since J is a finite dimensional linear space, the result of the preceding paragraph shows that V is finite. Put

$$p = \textstyle\sum_{M \in V} e_M.$$

Then p is well-defined and we shall show that p has the desired properties.

From (1) it is clear that p is a non-zero idempotent in A. Further, we shall prove that

$$\{p\}_r \cap J = (0). \tag{2}$$

Observe that $I = \{p\}_r \cap J$ is a closed ideal. If $I \neq (0)$, then, by Lemma II.2.1, I contains a minimal closed ideal N, say. Note that $N \in V$. Thus

$$pe_N = e_N \neq 0,$$

contradicting the fact that $e_N \in I \subset \{p\}_r$. Hence $I = (0)$, and (2) holds.

Next we show that

$$J = Ap. \tag{3}$$

Since $p \in J$, it is clear that $Ap \subset J$. On the other hand, if $x \in J$, then $-x \in J$ and thus $x - xp \in J$. But $x - xp$ is also contained in $\{p\}_r$. Hence we can use formula (2) to show that $x = xp$. This proves (3).

Now take $a \in A$. Then $ap \in J$, and hence $-ap \in J \subset A$. This implies that $a - ap \in A$. Clearly, $a - ap$ belongs to $\{p\}_r$. Since $a = (a - ap) + ap$, it follows that a is the sum of an element in $\{p\}_r$ and an element in Ap. Suppose

$$a = x_1 + x_2, \tag{4}$$

with $x_1 \in \{p\}_r$ and $x_2 \in Ap$. Then $px_1 = 0$, and thus $x_2 = ap$. But then $x_1 = a - ap$. This shows that the representation (4) is unique. Hence (i) holds. The conditions (ii) and (iii) are trivial consequences of formulas (2) and (3) and the fact that $J = A \cap (-A)$.

The proof of the sufficiency of the conditions is straightforward and is, therefore, omitted.

The next two lemmas are basic steps towards the proof of the characterization theorem of this section. The second lemma is a special case of this theorem.

3.2 LEMMA. *Let* t *be an element in* B*, and suppose that* A(t) *is locally compact and semi-simple. Then* Perσ(t) *is a set of simple poles of* t *and the associated spectral idempotent belongs to* A(t).

PROOF. Since A(t) is locally compact, $t \neq 0$. According to Proposition 1.3, this implies that $r(t) > 0$. Without loss of generality we may suppose

$r(t) = 1$. But then it follows that

$$S(t) = \mathrm{cl}\{t, t^2, \ldots\}$$

is a compact semi-group in B (see the paragraph preceding Proposition 2.3), and we can apply Theorem I.2.4 to get the desired result.

3.3 LEMMA. *Let* t *be a non-zero element in* B. *Then* A(t) *is locally compact, semi-simple and*

$$A(t) = -A(t)$$

if and only if $\sigma(t)$ *is a finite set of simple poles of* t *such that*

$$\sigma(t) \cap \mathbb{R}^+ = \emptyset.$$

PROOF. Suppose that A(t) is locally compact and semi-simple, and let $A(t) = -A(t)$. From the preceding lemma we know that $\mathrm{Per}\sigma(t)$ is a set of simple poles of t and that the associated spectral idempotent q belongs to A(t). Since A(t) is not strict, Theorem 2.1 shows that r(t) does not belong to $\mathrm{Per}\sigma(t)$, i.e.,

$$\mathrm{Per}\sigma(t) \cap \mathbb{R}^+ = \emptyset.$$

Let $s = t - tq$. If $\lambda \in \sigma(t) \setminus \mathrm{Per}\sigma(t)$, then we know from spectral theory that $\lambda \in \sigma(s)$ and that λ is a simple pole of t if and only if λ is a simple pole of s (cf., Proposition P.5). Hence it suffices to prove that s has the desired spectral properties. There is nothing to prove if $s = 0$. Suppose therefore that $s \neq 0$.

Since $q \in A(t) = -A(t)$, $s = t - tq \in A(t)$. Hence A(s) is locally compact and semi-simple. Further, it is easily seen that $A(s) = -A(s)$. So we can repeat the argument for s. From

$$A(t) = -A(t), \quad A(s) = -A(s),$$

it follows that A(t) and A(s) are real linear spaces. Local compactness of A(t) implies that A(t) is finite dimensional. Since $q \notin A(s)$, the dimension of A(s) is strictly less then the dimension of A(t). Hence the proof is completed by a finite number of repetitions of the foregoing argument.

Conversely, suppose that t has the described spectral properties. Theorem II.1.1 immediately implies that A(t) is locally compact. The element t can be written in the form

$$t = \lambda_1 e_1 + \ldots + \lambda_r e_r,$$

where $\{\lambda_1,\dots,\lambda_r\}$ is the non-zero part of $\sigma(t)$ and e_j is the spectral idempotent corresponding to the spectral set $\{\lambda_j\}$ $(j = 1,\dots,r)$. This implies that each s in A(t) admits a representation of the form

$$s = \alpha_1 e_1 + \dots + \alpha_r e_r \tag{5}$$

with $\alpha_1,\dots,\alpha_r$ in C. If s is of the form (5), then

$$s^2 = \alpha_1^2 e_1 + \dots + \alpha_r^2 e_r.$$

Now $s^2 = 0$ implies that $\alpha_j = 0$ $(j = 1,\dots,r)$ and therefore $s = 0$. Hence, by Proposition 1.2, A(t) is semi-simple. It remains to show that A(t) = -A(t).

Since A(t) is locally compact and semi-simple, we can apply Theorem 3.1. Let p be an idempotent in A(t) having the properties described in Theorem 3.1(i)-(iii) with A = A(t). It suffices to show that $\{p\}_r = (0)$.

Clearly

$$A(t - tp) = \{p\}_r.$$

Suppose $\{p\}_r \neq (0)$. Then, according to condition (iii) of Theorem 3.1, A(t - tp) is locally compact, semi-simple and strict. By Theorem 2.1, this implies that

$$0 < r(t - tp) \in \sigma(t - tp). \tag{6}$$

Since $p \in A(t)$, p is the uniform limit of polynomials in t. Thus, by Proposition P.6, p and e - p are spectral idempotents. It follows too that

$$\sigma(t - tp) \subset \sigma(t) \cup \{0\}.$$

Combining this result with (6) gives

$$0 < r(t - tp) \in \sigma(t),$$

but this contradicts the fact that $\sigma(t) \cap \mathbb{R}^+ = \emptyset$. Thus $\{p\}_r = (0)$, and the proof is complete.

Suppose that A(t) is locally compact, semi-simple and A(t) = -A(t). Then, by Lemma 3.3, the non-zero part of $\sigma(t)$ is a spectral set of t. Let q be the associated spectral idempotent. A careful examination of the proof of that lemma shows that

$$q \in A(t) \tag{7}$$

and that

$$tq = t.$$

Hence q is a unit element for A(t). In Chapter IV we shall reprove (7) in a more general setting.

The next theorem is the main result of this section.

3.4 THEOREM. *Let* t *be a non-zero element in* B. *Then* A(t) *is locally compact and semi-simple if and only if* $\sigma(t)$ *decomposes into two disjoint closed subsets* σ_1 *and* σ_2 *such that*

(i) σ_1 *is a finite (possibly empty) set of simple poles of* t *and* $\sigma_1 \cap \mathbb{R}^+ = \emptyset$;

(ii) *either* σ_2 *is empty or there exists* $0 < \alpha \in \sigma_2$ *such that*

$$\sigma_2 = \sigma(t) \cap \{\lambda: |\lambda| \leq \alpha\}$$

and

$$\sigma_2 \cap \{\lambda: |\lambda| = \alpha\}$$

is a finite set of simple poles of t.

PROOF. Suppose A(t) is locally compact and semi-simple. Then there exists an idempotent p in A(t) satisfying the conditions (i)-(iii) of Theorem 3.1 with A = A(t). It is easily seen that

$$A(t - tp) = \{p\}_r, \quad A(tp) = A(t)p.$$

Suppose tp = 0. Then A(t) = A(t - tp) is locally compact, semi-simple and strict, and we can apply Theorem 2.1 to show that

$$\sigma_1 = \emptyset, \quad \sigma_2 = \sigma(t)$$

gives a decomposition of $\sigma(t)$ with the required properties.

Suppose t - tp = 0. Then A(t) = A(t)p, and hence A(t) = -A(t). But then we can apply Lemma 3.3 to show that

$$\sigma_1 = \sigma(t), \quad \sigma_2 = \emptyset$$

gives a decomposition of $\sigma(t)$ with the required properties.

Next suppose that tp and t - tp are non-zero. Since $p \in A(t)$, p is a spectral idempotent, and hence there exists a decomposition of $\sigma(t)$ into disjoint closed sets δ_1 and δ_2 such that

$$\sigma(tp) = \delta_1 \cup \{0\}, \quad \sigma(t - tp) = \delta_2 \cup \{0\}.$$

Further we know from Proposition P.5 that any non-zero simple pole of tp or of t - tp is a simple pole of t. Since $t - tp \neq 0$, Theorem 3.1(iii) implies that A(t - tp) is locally compact, semi-simple and strict. Hence, by Theorem 2.1,

$$0 < r(t - tp) \in \sigma(t - tp)$$

and

$$\sigma(t - tp) \cap \{\lambda: |\lambda| = r(t - tp)\}$$

is a finite set of simple poles of t - tp and hence of t. Since $tp \neq 0$, Theorem 3.1(ii) implies that A(tp) is locally compact, semi-simple and A(tp) = -A(tp). Hence, by Lemma 3.3, $\sigma(tp)$ is a finite set of simple poles of tp and

$$\sigma(tp) \cap \mathbb{R}^+ = \emptyset.$$

Put $\alpha = r(t - tp)$, and take

$$\sigma_1 = \{\lambda \in \sigma(t): |\lambda| > \alpha\}$$

and

$$\sigma_2 = \{\lambda \in \sigma(t): |\lambda| \leq \alpha\}.$$

Then $\alpha > 0$. From what we proved above, it follows that σ_1 and σ_2 have the desired properties.

Conversely, let σ_1 and σ_2 be a decomposition of $\sigma(t)$ satisfying the conditions (i) and (ii). It is an immediate consequence of Theorem II.1.1 that A(t) is locally compact. It remains to show that A(t) is semi-simple.

Suppose that $\sigma_1 = \emptyset$ or $\sigma_2 = \emptyset$. Then we can apply either Theorem 2.1 or Lemma 3.3 to show that A(t) is semi-simple. Therefore, suppose that σ_1 and σ_2 are both non-empty. The sets σ_1 and σ_2 are complementary spectral sets. Let p be the spectral idempotent associated with σ_1. Then e - p is the spectral idempotent associated with σ_2. For each x in A(t), we have $xp \in A(tp)$, $x(e - p) \in A(t - tp)$ and

$$x = xp + x(e - p).$$

If $x^2 = 0$, then $(xp)^2 = x^2p = 0$ and $(x(e - p))^2 = x^2(e - p) = 0$. Hence, in order to prove that A(t) is semi-simple, it suffices to show that A(tp) and A(t - tp) are semi-simple (cf. Proposition 1.2).

From spectral theory we know that

$$\sigma(tp) = \sigma_1 \cup \{0\}, \quad \sigma(t - tp) = \sigma_2 \cup \{0\}$$

and 0 is a simple pole of tp (Proposition P.5). Further, if $\lambda \neq 0$ is a simple pole of t and $\lambda \in \sigma(tp)$ or $\lambda \in \sigma(t - tp)$, then λ is a simple pole of tp or t - tp respectively. Thus $\sigma(tp)$ is a set of simple poles of tp and

$$\sigma(tp) \cap \mathbb{R}^+ = \emptyset.$$

Further, $0 < r(t - tp) \in \sigma(t - tp)$ and the peripheral spectrum of $t - tp$ is a set of simple poles of $t - tp$. Hence we can apply Lemma 3.3 and Theorem 2.1, respectively, to show that $A(tp)$ and $A(t - tp)$ are semi-simple. This completes the proof.

We conclude this chapter with a proposition about the number of idempotents in $A(t)$.

3.5 PROPOSITION. *The number of idempotents in a semi-simple locally compact monothetic semi-algebra is finite.*

PROOF. Let $t \in B$ be such that $A(t)$ is locally compact and semi-simple. Let p be an idempotent in $A(t)$ satisfying the conditions (i)-(iii) of Theorem 3.1 with $A = A(t)$. In this case we know that

$$A(t - tp) = \{p\}_r, \quad A(tp) = A(t)p.$$

Clearly, it suffices to show that $A(t - tp)$ and $A(tp)$ contain finitely many idempotents. For $A(t - tp)$ this is an immediate consequence of Corollary 2.10. To show that $A(tp)$ contains a finite number of idempotents, we observe that any idempotent in $A(tp)$ is a spectral idempotent of tp (Proposition P.6). According to Lemma 3.3, the spectrum of tp is a finite set. Hence the number of spectral idempotents of tp is finite, and, therefore, the same is true for the number of idempotents in $A(tp)$.

NOTES ON CHAPTER III

1. An algebraic structure theory for semi-simple locally compact (not necessarily monothetic) semi-algebras has been developed by Bonsall in [2]. Some of his results have been incorporated in section 3 of Chapter II and the preceding sections, others will be mentioned in these notes.

2. In the remarks following the proof of Theorem 2.7 we observed that the spectral radius is an additive function on a strict, semi-simple locally compact monothetic semi-algebra. It turns out that the spectral radius is also multiplicative on such a semi-algebra. To see this, let $A(t)$ be locally compact, semi-simple and strict. Without loss of generality we may suppose that $r(t) = 1$. Then, according to formula (6) of section 2,

$$se_o = r(s)e_o \quad (s \in A(t)),$$

where e_o is some non-zero idempotent. Now take s_1 and s_2 in A(t), then

$$r(s_1 s_2)e_o = s_1 s_2 e_o = r(s_2)r(s_1)e_o,$$

and hence, since $e_o \neq 0$, $r(s_1 s_2) = r(s_1)r(s_2)$.

3. In [2] Bonsall has introduced the notion of primeness. A semi-algebra A is said to be *prime* if $J_1 J_2 \neq (0)$ whenever J_1 and J_2 are non-zero closed two-sided ideals in A. If A is commutative, then A is prime if and only if A has no divisors of zero. Clearly, primeness is a stronger condition than semi-simplicity. However, for a strict, locally compact monothetic semi-algebra, the two conditions coincide. Indeed, let A(t) be strict and locally compact, and suppose that A(t) is semi-simple. Take s_1 and s_2 in A(t) such that $s_1 s_2 = 0$. Since (see the preceding note) the spectral radius is multiplicative on A(t), we have

$$0 = r(s_1 s_2) = r(s_1)r(s_2).$$

But in a commutative semi-simple semi-algebra non-zero elements have a non-zero spectral radius (see Proposition 1.3). Hence at least one of the factors s_1 and s_2 must be zero. Thus A(t) has no divisors of zero, and, therefore, A(t) is prime.

4. Bonsall has proved that a locally compact semi-algebra is prime if and only if it is semi-simple and has exactly one minimal closed two-sided ideal (see Theorem 9 in [2]). Hence, using the result of the preceding note, a strict, semi-simple locally compact monothetic semi-algebra has exactly one minimal closed ideal. This result is also proved in section 2 of this Chapter, where it was obtained by using the spectral properties of a generating element.

5. In section 2 we proved that the minimal closed ideal M in a strict, semi-simple locally compact monothetic semi-algebra is of the form

$$M = \mathbb{R}_o^+ e_o,$$

where e_o is a non-zero idempotent. This implies that the ideals (closed or not) contained in M are M and (0). Hence the minimal closed ideal M is a minimal ideal. This result holds in general. Bonsall has proved that in a semi-simple locally compact semi-algebra any minimal closed right (left) ideal is a minimal right (left) ideal (see Theorem 8 in [2]).

6. Let A be a semi-algebra. Recall that an idempotent p in A is called a right minimal idempotent of A if pA is a minimal closed right ideal in A. Similarly, p is said to be a left minimal idempotent of A if Ap is a minimal closed left ideal in A. It is interesting to observe that in a semi-simple locally compact semi-algebra there is no distinction between the two. The reason for this is the fact that in such a semi-algebra A the element p is a right minimal idempotent if and only if pAp is a division semi-algebra, and similarly for a left minimal idempotent (see [2], Theorem 4, cf., Theorem II.2.6).

7. We conclude with Bonsall's structure theorem for semi-simple commutative locally compact semi-algebras (see [2], Theorem 11). Let A be such a semi-algebra. Then the set of minimal idempotents of A is a finite non-empty set $e_1,\ldots,e_n$. Each ideal e_kA is a closed commutative division semi-algebra with unit element e_k, and $e_ke_j = 0 \quad (k \neq j)$. If also A is strict, then

$$e_kA = \mathbb{R}_o^+ e_k \quad (k = 1,\ldots,n).$$

CHAPTER IV

POSITIVE OPERATORS

In section 1 of this chapter we obtain a converse of Theorem II.1.2 which gave sufficient conditions that a non-zero element of a Banach algebra should generate a strict locally compact semi-algebra. In fact we show in Theorem 1.7 that if the peripheral spectrum of the generator is a set of poles and if the semi-algebra it generates is strict then the spectral radius is a pole of the generator of maximal order in the peripheral spectrum. The proof of this result is based on Proposition 1.4, which gives sufficient conditions for an element in a semi-algebra to have its spectral radius in its spectrum. These conditions are phrased in terms of ideals of the semi-algebra.

Proposition 1.4 and Theorem 1.7 are due to Kaashoek [17], [18], and they are used in section 2 to put the Krein-Rutman theorem (on the existence of a positive eigenvector for the spectral radius of a positive operator) into a semi-algebra context (Theorem 2.1). Further we show that the same methods yield the generalizations of the Krein-Rutman theorem which were proved by Raghavan [27] and Sasser [30] (Theorems 2.4 and 2.7).

We carry out a similar programme for irreducible positive operators in section 3, employing and extending results of Bonsall and Tomiuk [4]. This culminates in Theorem 3.5 which states that if a bounded irreducible positive linear operator on a Banach space has a peripheral spectrum consisting of poles of finite rank then the spectral radius is an eigenvalue and the associated eigenspace is one dimensional.

1. STRICT SEMI-ALGEBRAS

Throughout this section B will be a complex Banach algebra with unit e and t will be a non-zero element in B.

1.1 LEMMA. *Let* σ *be a spectral set of* t *such that*

$$\sigma \cap \{\alpha \in \mathbb{R}: \alpha \geq 0\} = \emptyset,$$

and let q *be the spectral idempotent associated with* σ. *Then*

$$-q \in A(tq).$$

PROOF. Put $t_1 = tq$. The spectrum of t_1 consists of the set σ and, possibly, the point 0 (Proposition P.5). So our hypothesis implies that

$$\sigma(t_1) \cap \mathbb{R}^+ = \emptyset.$$

For $\lambda > r(t_1)$ we have

$$R(\lambda;t_1) = \sum_{n=1}^{\infty} \frac{1}{\lambda^n} t_1^{n-1}. \tag{1}$$

Since the Neumann series (1) converges in the norm of B, we have

$$t_1R(\lambda;t_1)^n \in A(t_1) \tag{2}$$

for $\lambda > r(t_1)$ and each n in $\mathbb{Z}^+$. Now

$$R(\mu;t_1) = \sum_{n=1}^{\infty} (\lambda - \mu)^{n-1} R(\lambda;t_1)^n$$

for μ close to λ. Therefore (2) implies that

$$t_1R(\mu;t_1) \in A(t_1) \tag{3}$$

for $0 < \mu < \lambda$ and μ close to λ. Since $\sigma(t_1) \cap \mathbb{R}^+ = \emptyset$, we can use a chain argument to prove that (3) holds for all $\mu > 0$.

Note that either 0 is a simple pole of t_1 with spectral idempotent $e - q$ or $0 \notin \sigma(t_1)$ and then $e - q = 0$. In both cases $t_1(e - q) = 0$ and $R(\mu;t_1)$ admits a Laurent expansion at 0 of the form

$$R(\mu;t_1) = \sum_{n=0}^{\infty} \mu^n a_n + \frac{1}{\mu}(e - q)$$

(cf., Proposition P.5). Now, by Proposition P.7(ii), we have $t_1a_o = -q$ and, since

$$t_1R(\mu;t_1) = \sum_{n=0}^{\infty} \mu^n t_1 a_n$$

for $\mu \neq 0$ and μ sufficiently small, we obtain

$$-q = t_1 a_o = \lim_{\mu \downarrow o} t_1 R(\mu; t_1) \in A(t_1).$$

This completes the proof.

Suppose that $\sigma(t)$ is a finite set of simple poles of t such that

$$\sigma(t) \cap \mathbb{R}^+ = \emptyset.$$

(These conditions occur in Lemma III.3.3). Then the non-zero part σ of $\sigma(t)$ is a spectral set of t satisfying the hypotheses of the preceding lemma. Let q be the spectral idempotent corresponding to σ. In this particular case $t(e - q) = 0$, and thus $tq = t$. But then the result of the preceding lemma implies that

$$-q \in A(t), \tag{4}$$

and hence

$$q = (-q)^2 \in A(t)$$

(this is formula (7) of section III.3). From $tq = t$ it also follows that q is a unit element of A(t). By combining this with formula (4), we obtain

$$A(t) = -A(t).$$

Hence the preceding lemma can be used to furnish an alternative for the last part of the proof of Lemma III.3.3.

Throughout the remainder of this section (with the exception of Proposition 1.8) we shall suppose that t is a non-zero element of B such that the peripheral spectrum of t, $\mathrm{Per}\sigma(t)$, is a set of poles of t. For such an element t, $\mathrm{Per}\sigma(t)$ is a spectral set of t. The associated spectral idempotent will be denoted by p.

1.2 PROPOSITION. *Suppose that* $r(t) \notin \sigma(t)$. *Then*

$$-p \in A(tp).$$

PROOF. The condition $r(t) \notin \sigma(t)$ implies $r(t) > 0$. But then $\mathrm{Per}\sigma(t)$ is a spectral set of t satisfying the hypotheses of Lemma 1.1, whence the result.

Suppose that $r(t) > 0$. Then $t^n \neq 0$ for each n in $\mathbb{Z}^+$. Denote by C(t) the set of cluster points of the sequence

$$\{\|t^n\|^{-1}t^n: n = 1,2,\ldots\}.$$

An element $h \in C(t)$ if, and only if, there exists a strictly increasing sequence $\{n_i\}$ of natural numbers such that

$$h = \lim_i \|t^{n_i}\|^{-1}t^{n_i}.$$

Observe that $C(\alpha t) = C(t)$ for each $\alpha > 0$.

1.3 PROPOSITION. *Suppose that* $r(t) > 0$. *Then the set* $C(t)$ *is non-empty, each* h *in* $C(t)$ *has norm one and*

$$ph = hp = h \quad (h \in C(t)).$$

PROOF. Without loss of generality we may suppose that $r(t) = 1$. Then $\|t^n\| \geq 1$ and hence $\|t^n\|^{-1} \leq 1$ for $n = 1,2,\ldots$.

From the definition of p it follows that $r\{t(e - p)\}$ is strictly less than one (cf., Proposition P.4(ii)). This implies that

$$t^n(e - p) = \{t(e - p)\}^n \to 0 \quad (n \to +\infty),$$

and, therefore,

$$(\|t^n\|^{-1}t^n)(e - p) \to 0 \quad (n \to +\infty). \tag{5}$$

Next we consider $t^n p$. From the spectral properties of t it follows that tp generates a finite dimensional subalgebra of B, B_0 say. Since $t^n p = (tp)^n$ for each n,

$$\{\|t^n\|^{-1}t^n p: n = 1,2,\ldots\}$$

is a sequence in B_0. Clearly this sequence is bounded. Hence there exists a strictly increasing sequence $\{n_i\}$ of natural numbers such that

$$\lim_i \|t^{n_i}\|^{-1}t^{n_i}p$$

exists. Combining this with the result of the preceding paragraph, we see that

$$\lim_i \|t^{n_i}\|^{-1}t^{n_i}$$

exists. Hence $C(t)$ is non-empty.

Take h in $C(t)$. Clearly, $\|h\| = 1$. From formula (5) it follows that $h(e - p) = 0$, and hence $h = hp$. Since t and p commute, we also have that h and p commute. Thus $ph = hp = h$.

1.4 PROPOSITION. *Suppose that* $r(t) > 0$, *and let* J *be a closed ideal in* $A(t)$ *such that*

(i) $J \cap (-J) = (0)$,

(ii) $C(t) \cap J \neq \emptyset$.

Then $r(t) \in \sigma(t)$.

PROOF. Suppose $r(t) \notin \sigma(t)$. Then we know from Proposition 1.2 that there exists a sequence $\{p_n\}$ of polynomials with non-negative coefficients such that

$$-p = \lim_n p_n(tp).$$

Take $h \in C(t) \cap J$. Then

$$-h = -ph = \lim_n p_n(tp)h.$$

Now h absorbs p, therefore $p_n(tp)h = p_n(t)h$, and hence, since J is an ideal,

$$p_n(tp)h \in J$$

for each n. But J is closed. So we obtain that $-h \in J$. Thus h belongs to J and -J. Since $J \cap (-J) = (0)$, this implies $h = 0$, contradicting the fact that $\|h\| = 1$. Hence $r(t) \in \sigma(t)$.

Suppose, in addition to the hypotheses on t, that A(t) is strict. Then by taking J in the preceding proposition to be A(t), we obtain $r(t) \in \sigma(t)$, provided that r(t) is strictly positive. But if $r(t) = 0$, then $r(t) \in \sigma(t)$ trivially. Hence to get a converse of Theorem II.1.2, it only remains to show that the pole r(t) is of maximal order in $\mathrm{Per}\sigma(t)$. The proof of this is based on the next two lemmas and the fact that, in this case, $r(t) \in \sigma(t)$ implies that A(t) is locally compact (see Theorem II.1.1).

1.5 LEMMA. *Let* A *be a strict locally compact semi-algebra in* B. *Then there exists* $\gamma \geq 1$ *such that*

$$\|x\| \leq \gamma\|x + y\| \quad (x,y \in A).$$

PROOF. Let $\delta = \inf \{\|x + y\|: x,y \in A, \|x\| = 1\}$. Suppose that $\delta = 0$. Then there exist sequences $\{x_n\}$ and $\{y_n\}$ in A such that

$$\lim (x_n + y_n) = 0, \quad \|x_n\| = 1 \quad (n = 1,2,\ldots).$$

Since A is locally compact, the sequence $\{x_n\}$ has a converging subsequence.

Passing to this subsequence, we may suppose that $\lim x_n = x_o$ exists. But then

$$\lim y_n = \lim \{(x_n + y_n) - x_n\} = -x_o.$$

Since A is closed, this implies that $x_o \in A \cap (-A)$, and hence $x_o = 0$. But this contradicts the fact that $\|x_o\| = \lim \|x_n\| = 1$. Thus $\delta > 0$. From the definition it follows that $\delta \leq 1$. Hence we can take $\gamma = \delta^{-1}$.

1.6 LEMMA. *Let* A *be a strict locally compact semi-algebra in* B, *and let* A_e *be the smallest closed semi-algebra in* B *containing* A *and the unit element* e. *Then* A_e *is strict and locally compact.*

PROOF. If $e \in A$, then there is nothing to prove. Therefore, suppose that $e \notin A$. Then it is easy to see that

$$A_e = \{a + \alpha e : a \in A, \alpha \geq 0\}.$$

Using this representation for A_e it is not difficult (though somewhat tedious) to verify that A_e has the required properties.

We now come to the main theorem of this section.

1.7 THEOREM. *Let* t *be a non-zero element of* B *such that* $\mathrm{Per}\sigma(t)$ *is a set of poles of* t. *Suppose that* A(t) *is strict. Then* r(t) *is a pole of* t *of maximal order in* $\mathrm{Per}\sigma(t)$.

PROOF. If $r(t) = 0$, then the theorem is trivially true. Therefore, suppose that $r(t) > 0$. Then we know from Proposition 1.4 that $r(t) \in \sigma(t)$. Hence r(t) is a pole of t, and it remains to show that it is of maximal order in $\mathrm{Per}\sigma(t)$.

Since $r(t) \in \sigma(t)$, we can apply Theorem II.1.1 to show that A(t) is locally compact. Let A be the smallest closed semi-algebra in B containing A(t) and the unit element e. Then it follows from Lemma 1.6 and the hypotheses on t that A is strict and locally compact. Hence, by Lemma 1.5, there exists $\gamma \geq 1$ such that

$$\|x\| \leq \gamma \|x + y\| \quad (x, y \in A). \tag{6}$$

Let k be the order of the pole r(t), and choose an integer $m > k$. Further, let $\alpha > r(t)$. Consider

$$f(\alpha) = \{\alpha - r(t)\}^m(\alpha e - t)^{-1}.$$

Since $\alpha > r(t)$, we have

$$f(\alpha) = \{\alpha - r(t)\}^m\left(\sum_{n=0}^{\infty} \frac{1}{\alpha^{n+1}} t^n\right).$$

If $\{\delta_n\}$ is a sequence of real numbers such that

$$0 \leq \delta_n \leq 1 \quad (n = 1,2,\ldots), \tag{7}$$

then the series

$$\sum_{n=0}^{\infty} \frac{1}{z^{n+1}} \delta_n t^n$$

converges for $|z| > r(t)$. Therefore

$$g(\alpha) = \{\alpha - r(t)\}^m\left(\sum_{n=0}^{\infty} \frac{1}{\alpha^{n+1}} \delta_n t^n\right) \tag{8}$$

is well-defined for $\alpha > r(t)$. From condition (7) and the fact that $\alpha > r(t)$ we have

$$g(\alpha) \in A, \quad f(\alpha) - g(\alpha) \in A.$$

But then it follows from (6) that

$$\|g(\alpha)\| \leq \gamma\|g(\alpha) + \{f(\alpha) - g(\alpha)\}\| = \gamma\|f(\alpha)\|.$$

Since m is strictly greater than the order of the pole $r(t)$, we have

$$\lim_{\alpha\downarrow r(t)} f(\alpha) = 0,$$

thus

$$\lim_{\alpha\downarrow r(t)} g(\alpha) = 0. \tag{9}$$

A sequence of real numbers of absolute value less than or equal to one can be written as the difference of two sequences both satisfying condition (7). Hence (9) remains true if in (8) the sequence $\{\delta_n\}$ is a sequence of real numbers with $|\delta_n| \leq 1$ for each n.

Take $\lambda \neq r(t)$ in $\operatorname{Per}\sigma(t)$. Then $\lambda = r(t)\exp(-i\phi)$ for some ϕ $(0 < \phi < 2\pi)$. Put $z = \alpha \exp(-i\phi)$ with $\alpha > r(t)$. Since

$$(ze - t)^{-1} = \sum_{n=0}^{\infty} \frac{1}{\alpha^{n+1}} \{\cos(n+1)\phi + i \sin(n+1)\phi\}t^n,$$

the result of the preceding paragraph implies that

$$\lim_{\alpha\downarrow r(t)} \|(z - \lambda)^m(ze - t)^{-1}\| = 0.$$

Hence the order of the pole λ is strictly less than m. This holds for each

$m > k$. Thus the order of the pole λ is less than or equal to k, and hence the pole $r(t)$ is of maximal order in $\mathrm{Per}\sigma(t)$. This completes the proof.

Let t be a non-zero element in B such that $\mathrm{Per}\sigma(t)$ is a set of poles of t. By combining Theorem II.1.2 and the result of the preceding theorem, we obtain a necessary and sufficient condition for strictness of $A(t)$, namely, $A(t)$ is strict if and only if $r(t)$ is a pole of t of maximal order in $\mathrm{Per}\sigma(t)$.

Theorem 1.7 implies that there exists a non-zero element s in the smallest closed semi-algebra in B containing t and e such that

$$ts = r(t)s.$$

This follows from the following proposition.

1.8 PROPOSITION. *Let* t *be an element in* B *such that* $r(t)$ *is a pole of* t*, and let* A *be the smallest closed semi-algebra in* B *containing* t *and the unit element* e*. Then there exists* $s \neq 0$ *in* A *such that*

$$ts = r(t)s.$$

PROOF. Let k be the order of $r(t)$ as a pole of t. Then

$$s = \lim_{\lambda \downarrow r(t)} \{\lambda - r(t)\}^k (\lambda e - t)^{-1}$$

exists and $s \neq 0$. From the Neumann series for $(\lambda e - t)^{-1}$ it follows that $s \in A$. Further we know that

$$t(\lambda e - t)^{-1} = \lambda(\lambda e - t)^{-1} - e.$$

Using this in the definition of s, one obtains easily that $ts = r(t)s$.

2. THE KREIN-RUTMAN THEOREM AND SOME OF ITS GENERALIZATIONS

We begin this section with a few definitions concerning positive operators.

Let E be a (real or complex) linear space. A non-empty subset K of E is called a *cone* in E if K satisfies the following conditions

(i) $K + K \subset K$,

(ii) $\alpha K \subset K$ for each $\alpha \geq 0$,

(iii) $K \cap (-K) = \{0\}$.

Let K be a cone in E. The partial ordering induced by K in E will be denoted by $\leq$. Thus $x \leq y$ (or equivalently $y \geq x$) whenever $y - x \in K$. Observe that the cone axioms imply that the partial ordering $\leq$ is compatible with the linear structure of E. Further

$$K = \{x \in E: x \geq 0\}.$$

An *ordered Banach space* E is a Banach space with a closed cone K which induces the ordering on E. Often K is called the *positive cone* of E.

A linear operator T acting on an ordered Banach space E is called *positive* if T leaves invariant the cone of E. In other words, T is positive whenever

$$Tx \geq 0 \quad (x \geq 0).$$

In this section we shall, among other things, study the spectral properties of positive operators. Since the underlying space can be real, we have to define spectrum, spectral radius etc. for a linear operator acting on a real space.

Let E be a real Banach space. Denote by $E \oplus iE$ the complexification of E (see [31], p.33) endowed with the norm

$$\|x + iy\| = \sup\{\|x \cos \theta + y \sin \theta\|: 0 \leq \theta < 2\pi\}. \tag{1}$$

Then $E \oplus iE$ is a complex Banach space, $E \subset E \oplus iE$ and the norm (1) on E coincides with the original norm on E. Given a bounded linear operator T on E, we denote by T^c the operator on $E \oplus iE$ defined by

$$T^c(x + iy) = Tx + iTy.$$

It is easily seen that T^c is a bounded linear operator on $E \oplus iE$ and $\|T^c\| = \|T\|$. We call the operator T^c the *complexification* of T. Observe that the map

$$T \to T^c \tag{2}$$

is an isometric real algebra-isomorphism.

Let T be a bounded linear operator on the real Banach space E. Then we define the spectral properties of T to be those of its complexification T^c. Thus, by definition, $\sigma(T) = \sigma(T^c)$, $r(T) = r(T^c)$, $\text{Per}\sigma(T) = \text{Per}\sigma(T^c)$, poles of T are poles of T^c, etc.. Since the map (2) is an isometric homomorphism, we have

$$\|(T^c)^n\| = \|(T^n)^c\| = \|T^n\|$$

for each n. Hence it follows that the usual formula for the spectral radius holds, that is,

$$r(T) = \lim \|T^n\|^{1/n}.$$

Let A(T) denote the smallest closed semi-algebra in L(E) containing T, and let $A(T^c)$ be the smallest closed semi-algebra in $L(E \oplus iE)$ containing T^c. The map (2) induces an isometric (semi-algebra-) isomorphism of A(T) onto $A(T^c)$. Hence the algebraic and topological properties of A(T) and $A(T^c)$ are the same. Our goal is to obtain information about the spectral properties of T by examining A(T) and applying the results of the preceding sections. Since the spectral properties of T are those of T^c, and since A(T) and $A(T^c)$ are indistinguishable algebraically and topologically, we may, without loss of generality, identify T and T^c. This will be done in the sequel.

We now come to the first main theorem of this section. Recall that a subset V of a Banach space E is said to be *total* if the linear span of V is dense in E. If K is a cone in a real Banach space E, then K is total in E if and only if

$$E = \mathrm{cl}(K - K).$$

2.1 THEOREM. *Let* E *be a real or complex ordered Banach space with a total positive cone* K, *let* T *be a bounded positive linear operator on* E, *and suppose that the peripheral spectrum of* T, $\mathrm{Per}\sigma(T)$, *is a set of poles of* T. *Then*

(i) $r(T) \in \sigma(T)$;

(ii) *the order of the pole* r(T) *is maximal in* $\mathrm{Per}\sigma(T)$;

(iii) *there exists* $0 \neq u \in K$ *such that*

$$Tu = r(T)u.$$

PROOF. Consider the smallest closed semi-algebra A(T) generated by T. Since E is an ordered Banach space, the cone K is closed in E by definition, and hence all elements of A(T) are positive operators. Take

$$S \in A(T) \cap (-A(T)). \tag{3}$$

Then S and -S are positive operators, and thus S is zero on the cone K. But, since S is a bounded linear operator and since K is a total subset of E, this implies that S is the zero operator. Thus A(T) is strict, and we apply Theorem 1.7 to get (i) and (ii).

To prove (iii), observe that by (i) there exists a non-zero element S in the smallest closed semi-algebra A containing T and the identity operator on E such that

$$TS = r(T)S$$

(see Proposition 1.8). Since S is non-zero, the fact that K is a total subset of E implies the existence of an element v in K such that $u = Sv \neq 0$. But S is positive, thus $u \in K$ and

$$Tu = TSv = r(T)Sv = r(T)u.$$

This completes the proof of the theorem.

Let T be a compact positive linear operator acting on a real ordered Banach space E with a total positive cone K, and suppose that $r(T) > 0$. Then T satisfies the conditions of the preceding theorem (Proposition P.10) and therefore (i), (ii) and (iii) in 2.1 hold for these operators. This result is known as the Krein-Rutman theorem (see Theorem 6.1 in [22]).

In what follows we shall show that in Theorem 2.1 the positivity condition on the linear operator T can be weakened considerably without invalidating the statements (i), (ii) and (iii). For the first generalization (which is essentially due to D.W. Sasser [30]) we need the notion of a dual cone.

Let E be an ordered Banach space with cone K, and let E' denote the (topological) dual of E endowed with the dual norm. Then E' is a Banach space. Consider in E' the set

$$K' = \{f \in E': \ \mathrm{Re} f(u) \geq 0 \ (u \in K)\}.$$

It is easily seen that

$$K' + K' \subset K', \quad \alpha K' \subset K' \quad (\alpha \geq 0)$$

But it is not necessarily true that

$$K' \cap (-K') = (0), \tag{4}$$

that is, K' is not necessarily a cone in E'. In spite of this we call K' the *dual cone* of K in E'. If, in addition, K is total in E then (4) holds and, since K' is closed in E', the space E' with the ordering induced by K' is an ordered Banach space. Observe that the separation theorems for convex sets (see [31], section II.9) imply that

$$K = \{x \in E: \ \mathrm{Re} f(x) \geq 0 \ (f \in K')\}. \tag{5}$$

Let T be a bounded linear operator on the ordered Banach space E, and let K be the positive cone of E. We define T to be *quasi-positive* if for each u in K and each f in K', there exists a positive integer n(u,f) such that

$$\operatorname{Re} f(T^n u) \geq 0 \quad (n \geq n(u,f)).$$

We shall show that Theorem 2.1 holds for quasi-positive operators. In order to do this we need two lemmas.

Let T be a bounded linear operator on a Banach space E. For each positive integer n define $A_n(T)$ to be the smallest closed semi-algebra in L(E) containing the set

$$\{T^n, T^{n+1}, \ldots\}.$$

Put

$$A_\infty(T) = \bigcap_{n=1}^{\infty} A_n(T).$$

Observe that $A_1(T) = A(T)$. Further, it is easily seen that $A_n(T)$ $(n = 1,2,\ldots)$ and $A_\infty(T)$ are closed ideals in A(T). In the case that T is a quasi-positive operator acting on an ordered Banach space with a total cone the semi-algebra $A_\infty(T)$ is an interesting object. Because then we can show that all elements of $A_\infty(T)$ are positive operators and hence, since the cone is total, it will follow that $A_\infty(T)$ is strict.

2.2 LEMMA. *Let* T *be a bounded linear operator on a real or complex Banach space. Then the following statements are equivalent.*

(i) A(T) *is strict*;

(ii) $A_n(T)$ *is strict for each* n;

(iii) $A_n(T)$ *is strict for some* n.

PROOF. The implications (i) ⇒ (ii) and (ii) ⇒ (iii) are trivially true. To prove (iii) ⇒ (i), we first of all observe that A(T) is strict if T is nilpotent (see Theorem II.1.2). Hence, without loss of generality, we may suppose that $T^k \neq 0$ for $k = 1,2,\ldots$. Now let $A_n(T)$ be strict for some fixed $n \geq 2$. Consider the set

$$H = \{\alpha T^{n-1} + S: \alpha \geq 0,\ S \in A_n(T)\}.$$

Using the non-nilpotency of T and the fact that $A_n(T)$ is a strict closed ideal in A(T), it is not difficult to prove that H is a strict closed semi-algebra. Clearly, $A_{n-1}(T) \subset H$, and therefore $A_{n-1}(T)$ is strict. By repeating the argument we obtain after a finite number of steps that $A(T) = A_1(T)$ is strict. This completes the proof.

2.3 LEMMA. *Let* T *be a bounded linear operator on a real or complex Banach space, and suppose that* $A(T)$ *is locally compact. Then the following statements are equivalent.*

(i) $A(T)$ *is strict;*

(ii) $A_\infty(T)$ *is strict.*

PROOF. The implication (i) $\Rightarrow$ (ii) is trivial. To prove (ii) $\Rightarrow$ (i), suppose that $A(T)$ is not strict. Then, by the preceding lemma, $A_n(T)$ is not strict for any n. Hence, for $n = 1,2,\ldots$, there exists

$$x_n \in A_n(T) \cap -A_n(T)$$

with $\|x_n\| = 1$. Since $A(T)$ is locally compact, the sequence $\{x_n\}$ has a converging subsequence with limit y, say. Clearly

$$0 \neq y \in A_n(T) \cap -A_n(T)$$

for each n. But then y is a non-zero element in $A_\infty(T) \cap -A_\infty(T)$, contradicting the fact that $A_\infty(T)$ is strict. Hence $A(T)$ is strict, and the proof of the lemma is complete.

2.4 THEOREM. *Let* E *be a real or complex ordered Banach space with a total positive cone* K, *let* T *be a bounded quasi-positive linear operator on* E, *and suppose that the peripheral spectrum of* T *is a set of poles of* T. *Then*

(i) $r(T) \in \sigma(T)$;

(ii) *the order of the pole* $r(T)$ *is maximal in* $\mathrm{Per}\sigma(T)$;

(iii) *if* $r(T) > 0$, *there exists* $0 \neq u \in K$ *such that*

$$Tu = r(T)u.$$

PROOF. If $r(T) = 0$, then (i) and (ii) hold trivially. Therefore suppose that $r(T) > 0$. Observe that $A_\infty(T)$ is a closed ideal in $A(T)$. We shall prove (i) by showing that $A_\infty(T)$ satisfies the conditions on J in Proposition 1.4.

Let $C(T)$ be the set of cluster points of the sequence

$$\{\|T^n\|^{-1}T^n : n = 1,2,\ldots\}.$$

Clearly, $C(T) \subset A_n(T)$ for each n, and therefore $C(T) \subset A_\infty(T)$. According to

Proposition 1.3, the set C(T) is non-empty. So we have

$$C(T) = C(T) \cap A_\infty(T) \neq \emptyset. \tag{6}$$

To prove that $A_\infty(T)$ is strict, let w be an element in K and take f in K'. Since T is quasi-positive, there exists a positive integer $k = n(w,f)$ such that

$$\mathrm{Re}(f(T^n w)) \geq 0 \quad (n \geq k).$$

This implies that for each element S of the form

$$S = \alpha_k T^k + \alpha_{k+1} T^{k+1} + \ldots + \alpha_{k+r} T^{k+r} \tag{7}$$

with $\alpha_{k+j} \geq 0$ $(j = 0,1,\ldots,r)$, we have

$$\mathrm{Re} f(Sw) \geq 0. \tag{8}$$

The semi-algebra $A_k(T)$ is the closure in L(E) of the set of all operators S of the form (7) with non-negative coefficients $\alpha_k,\ldots,\alpha_{k+r}$. Therefore (8) holds for any element in $A_k(T)$. In particular, since $A_\infty(T)$ is a subset of $A_k(T)$, it follows that (8) holds for any element S in $A_\infty(T)$. So given w in K and f in K', we have

$$\mathrm{Re} f(Sw) \geq 0 \quad (S \in A_\infty(T)),$$

and thus we can use formula (5) to show that for each S in $A_\infty(T)$

$$Sw \geq 0 \quad (w \in K).$$

Hence $A_\infty(T)$ is a set of positive operators. But then, since K is a total cone, the arguments of the first part of the proof of Theorem 2.1 show that

$$A_\infty(T) \cap -A_\infty(T) = (0). \tag{9}$$

By combining this with (6), we see that $A_\infty(T)$ satisfies the conditions of J in Proposition 1.4. Thus $r(T) \in \sigma(T)$.

Next we prove (ii). The fact that $\mathrm{Per}\sigma(T)$ is a set of poles of T together with the final result of the preceding paragraph implies that A(T) is locally compact (Theorem II.1.1). Then, by Lemma 2.3, it follows from (9) that A(T) is strict. Thus T satisfies the conditions of Theorem 1.7, and, therefore, (ii) holds.

To prove (iii), let $A_0(T)$ be the smallest closed semi-algebra in L(E) containing T and the identity operator on E. It is easily seen that

$$\{T^n U \colon U \in A_0(T)\} \subset A_n(T) \tag{10}$$

for each n. Since (i) holds, we know from Proposition 1.8 that there exists a non-zero S in $A_o(T)$ such that

$$TS = r(T)S,$$

and hence for each n

$$T^n S = r(T^n)S.$$

But then, using formula (10) and the fact that $r(T) > 0$, it follows that S belongs to $A_n(T)$ for $n = 1,2,\ldots$. Thus $S \in A_\infty(T)$, and we may conclude that S is a positive operator. Proceeding as in the last part of the proof of Theorem 2.1, one shows that this implies the existence of a non-zero element u in K such that $Tu = r(T)u$. This completes the proof of the theorem.

Without the additional condition $r(t) > 0$ statement (iii) in Theorem 2.4 does not hold. To see this take E to be $\mathbb{R}^2$, let K be the set

$$\{(x,y) \in \mathbb{R}^2 : \ |y| \leq x\},$$

and let T be given by the matrix

$$T = \begin{pmatrix} 0 & 0 \\ -1 & 0 \end{pmatrix}.$$

Then K is a closed and total cone in $\mathbb{R}^2$. Further T^2 is the zero operator, and hence T satisfies the conditions of Theorem 2.4. Observe that $r(T) = 0$. In this case, $T(x,y) = 0$ implies that $x = 0$. Thus there does not exist a non-zero u in K such that $Tu = r(T)u$.

One may consider Theorem 2.1 to be a statement about the spectral properties of linear operators leaving invariant certain rather specific closed convex sets, namely closed cones. We conclude this section by showing that one can obtain similar results for operators which leave invariant other types of convex sets (cf. T.E.S. Raghavan [27]).

Let E be a real or complex Banach space, and let C be a closed convex set in E such that

$$C \cap -C = (0).$$

Thus 0 belongs to C and 0 is an extreme point of C. In Theorem 2.1 we need closed cones. Therefore it is important to note that the cone K generated by C may not be closed and that the closure $\overline{K}$ of K is not necessarily a cone. To see this, we consider the following example which is due to Raghavan (see [27]). Let E be $\mathbb{R}^2$, and take

$$C = \{(x,y): x \geq y^2\}.$$

Then the cone K generated by C does not contain the non-zero points on the y-axis. But these points are in the closure $\overline{K}$ of K. Hence in this case K is not closed, and, since $\overline{K} \cap -\overline{K}$ contains non-zero points, $\overline{K}$ is not a cone.

Let T be a bounded linear operator on E such that C is invariant under T, that is,

$$TC \subset C.$$

Then

$$T^n C \subset C$$

for each n, and hence, since $0 \in C$, the convex set C is invariant under all operators S of the form

$$S = \alpha_1 T + \alpha_2 T^2 + \ldots + \alpha_r T^r$$

with $\alpha_i \geq 0$ $(i = 1,\ldots,r)$ and $\sum_{i=1}^r \alpha_i \leq 1$. For an arbitrary bounded linear operator T on E, let the symbol D(T) denote the closure in L(E) of the set of all operators S of the form described above. In other words, let D(T) be the closed convex hull in L(E) of the set

$$\{0,T,T^2,\ldots\}.$$

Since C is a closed convex set containing 0, the foregoing implies that C is invariant under T if and only if

$$SC \subset C \quad (S \in D(T)). \tag{11}$$

Observe that D(T) is a convex multiplicative semigroup containing the zero operator 0.

The next two lemmas play an important role in the proof of the last theorem of this section.

2.5 LEMMA. *Let* T *be a bounded linear operator on a real or complex Banach space, and let*

$$D_\infty(T) = \{S \in D(T): \alpha S \in D(T)\ (\alpha > 0)\}.$$

Then $D_\infty(T)$ *is a closed ideal in* A(T).

PROOF. Clearly, $D_\infty(T) \subset D(T) \subset A(T)$. The fact that D(T) is closed implies that $D_\infty(T)$ is a closed subset of A(T). Further, since D(T) is a convex multiplicative semigroup containing 0, $D_\infty(T)$ is a semi-algebra.

Hence it remains to show that

$$SS_1 \in D_\infty(T) \quad (S \in A(T),\ S_1 \in D_\infty(T)). \tag{12}$$

To prove this, we first of all suppose that S is of the form

$$S = \alpha_1 T + \ldots + \alpha_r T^r \tag{13}$$

with $\alpha_i \geq 0$ $(i = 1,\ldots,r)$. Take $\beta > (\alpha_1 + \ldots + \alpha_r)$, and let α be an arbitrary positive number. Then $\beta^{-1}S$ and $\alpha\beta S_1$ are in D(T). But D(T) is closed under multiplication. Thus

$$\alpha SS_1 = (\beta^{-1}S)(\alpha\beta S_1) \in D(T).$$

This holds for each $\alpha > 0$. Therefore, $SS_1 \in D_\infty(T)$. This result together with the fact that $D_\infty(T)$ is a closed set and the fact that A(T) is the closure of the set of operators S of the form (13) with $\alpha_i \geq 0$, implies (12). This completes the proof.

2.6 LEMMA. *Let* T *be a bounded linear operator on a real or complex Banach space, and suppose that* $1 \leq r(T) \in \sigma(T)$. *Then*

$$A(T) = \{\alpha S\colon \alpha > 0,\ S \in D(T)\}.$$

PROOF. Clearly the right hand side is a subset of A(T). To prove the reverse inclusion, let S be of the form (13) with $\alpha_i \geq 0$ for $i = 1,\ldots,r$. Since $r(T) \in \sigma(T)$, we can use Proposition P.4(ii) to show that

$$r(S) = \textstyle\sum_{i=1}^{r} \alpha_i \{r(T)\}^i.$$

Now $r(T) \geq 1$, thus $r(S) \geq (\alpha_1 + \ldots + \alpha_r)$, and hence

$$\{1 + r(S)\}^{-1} S \in D(T). \tag{14}$$

Since D(T) is closed, the continuity of the spectral radius on commutative sets (Proposition P.2(iii)), implies that (14) holds for each operator in the closure of the set of operators S of the form (13) with $\alpha_i \geq 0$. Thus (14) holds for any S in A(T), and the proof of the lemma is complete.

2.7 THEOREM. *Let* E *be a real or complex Banach space, let* C *be a closed convex and total subset of* E *such that*

$$C \cap -C = (0),$$

and let T *be a bounded linear operator on* E *leaving* C *invariant. Suppose that* $r(T) \geq 1$ *and that the peripheral spectrum of* T *is a set of poles of* T. *Then*

(i) $r(T) \in \sigma(T)$;

(ii) *the order of the pole* $r(T)$ *is maximal in* $\mathrm{Per}\sigma(T)$;

(iii) *there exists* $0 \neq u \in C$ *such that*

$$Tu = r(T)u.$$

PROOF. The proof consists of four parts. 1. We begin by showing that

$$D(T) \cap -D(T) = \{0\}. \tag{15}$$

Take S in $D(T) \cap -D(T)$. Then S and -S belong to D(T), and we can use formula (11) to show that

$$SC \subset \{C \cap -C\}.$$

According to our hypotheses the right hand side of the last formula consists of the zero element only. Thus S is zero on C, and, since C is a total subset of E, this implies that S = 0. Hence (15) holds.

2. Suppose, in addition to our hypotheses, that T has equibounded iterates, i.e., there exists M > 0 such that

$$\|T^n\| \leq M \quad (n = 1,2,\ldots).$$

This together with our hypotheses on the spectrum of T implies that $r(T) = 1$ and the peripheral spectrum of T is a set of simple poles of T (Proposition P.9). But then we can apply Theorem I.2.3 to show that the monothetic semigroup S(T) generated by T is compact.

Let Σ denote the closed convex hull of S(T). Then the result of the preceding paragraph implies that Σ is compact (cf., [9], Theorem V.2.6). Further $\Sigma \subset D(T)$. We shall prove that $0 \notin \Sigma$.

Suppose $0 \in \Sigma$. Since $\Sigma \subset D(T)$, it follows from formula (15) that

$$\Sigma \cap -\Sigma = \{0\}.$$

In other words 0 is an extreme point of Σ. So we can use Milman's converse of the Krein-Milman theorem ([9], Lemma V.8.5) to show that $0 \in S(T)$. But we know that S(T) contains only one idempotent (Theorem I.1.1(iii)) and this idempotent (or its complexification) is the spectral idempotent associated with the spectral set $\mathrm{Per}\sigma(T)$ (Theorem I.2.4). Since $\mathrm{Per}\sigma(T) \neq \emptyset$, the sole idempotent of S(T) must be non-zero. Contradiction. Thus $0 \notin \Sigma$.

Take $S \neq 0$ in A(T). Then there exists a sequence of non-zero polynomials p_n with non-negative coefficients and first coefficient 0 such that

$$p_n(T) \to S \quad (n \to +\infty).$$

If $\sup_n p_n(1) = +\infty$, then there exists a strictly increasing sequence $\{n_i\}$ of positive integers such that $p_{n_i}(1) \to +\infty$ $(i \to +\infty)$ and hence

$$\{p_{n_i}(1)\}^{-1} p_{n_i}(T) \to 0 \quad (i \to +\infty).$$

But $\{p_n(1)\}^{-1} p_n(T) \in \Sigma$. Since Σ is closed, this implies $0 \in \Sigma$. Contradiction. Therefore there exists a positive number β such that

$$p_n(1) \leq \beta < +\infty \quad (n = 1,2,\ldots).$$

But then

$$\frac{1}{\beta} S = \lim_n \frac{1}{\beta} p_n(T) \in D(T),$$

and hence

$$A(T) = \{\alpha S\colon \alpha > 0,\ S \in D(T)\}. \tag{16}$$

Since $D(T) \cap -D(T) = \{0\}$, it follows that $A(T) \cap -A(T) = \{0\}$. In other words, $A(T)$ is strict, and we can use Theorem 1.7 to prove (i) and (ii).

3. Next we suppose that T does not have equibounded iterates. Consider the set $C(T)$ of cluster points of the sequence

$$\{\|T^n\|^{-1} T^n\colon n = 1,2,\ldots\}.$$

Firstly we prove that $C(T) \cap D_\infty(T) \neq \emptyset$.

Since T does not have equibounded iterates, there exists a strictly increasing sequence $\{n_i\}$ of positive integers such that

$$\|T^{n_i}\| \to +\infty \quad (i \to +\infty).$$

Repeating the arguments of the first part of the proof of Proposition 1.3, we see that the sequence

$$\{\|T^{n_i}\|^{-1} T^{n_i}\colon i = 1,2,\ldots\}$$

has a cluster point, Q say. Hence, by passing to a subsequence, we may assume that

$$\|T^{n_i}\|^{-1} T^{n_i} \to Q \quad (i \to +\infty).$$

Take $\alpha > 0$. For i sufficiently large we have $\|T^{n_i}\| \geq \alpha$. Hence, using the convexity of $D(T)$, we obtain that

$$\alpha \|T^{n_i}\|^{-1} T^{n_i} \in D(T)$$

for i sufficiently large. Thus $\alpha Q \in D(T)$. This holds for each $\alpha > 0$. Hence $Q \in D_\infty(T)$, and it follows that

$$C(T) \cap D_\infty(T) \neq \emptyset.$$

From formula (15) and $D_\infty(T) \subset D(T)$ it follows that

$$D_\infty(T) \cap (-D_\infty(T)) = \{0\}.$$

Thus $D_\infty(T)$ is a closed ideal in $A(T)$ (see Lemma 2.5) satisfying the conditions (i) and (ii) on J in Proposition 1.4. Thus $r(T) \in \sigma(T)$.

The result of the preceding paragraph together with the hypothesis $r(T) \geq 1$ shows (see Lemma 2.6) that formula (16) holds. But then, as in part 2 of the proof, it follows that (i) and (ii) hold.

4. It remains to prove (iii). By Proposition 1.8 there exists a non-zero operator S in the smallest closed semi-algebra generated by T and the identity operator on E such that

$$TS = r(T)S.$$

In fact, since $r(T) \geq 1$,

$$S = \{r(T)\}^{-1}TS \in A(T).$$

By multiplying S with a suitable positive constant, we may assume that $S \in D(T)$ (see formula (16)). Since C is a total subset of E, the fact that S is non-zero implies the existence of an element v in C such that $Sv = u \neq 0$. Formula (11) shows that $u \in C$. Further we have

$$Tu = TSv = r(T)Sv = r(T)u.$$

This completes the proof of the theorem.

If the convex set C in Theorem 2.7 is a cone, then the condition $r(T) \geq 1$ is superfluous. For then the result follows from Theorem 2.1. On the other hand if T is a positive operator acting on an ordered Banach space E, then any positive multiple αT of T leaves invariant the cone K of E and

$$r(\alpha T) = \alpha r(T).$$

Thus, if $r(T) > 0$ and $\alpha \geq \{r(T)\}^{-1}$, we obtain a linear operator T_1 with $r(T_1) \geq 1$ such that K is invariant under T_1. This shows that, for an operator T with $r(T) > 0$, Theorem 2.1 is a special case of the preceding theorem.

In general the condition $r(T) \geq 1$ in Theorem 2.7 may not be replaced by $r(T) > 0$. For example, let E be $\mathbb{R}^2$, let

$$C = \{(x,y): x \geq y^2\},$$

and let T be given by the matrix

$$\begin{pmatrix} \beta^2 & 0 \\ 0 & -\beta \end{pmatrix}$$

with $0 < \beta < 1$. Then C is invariant under T, $r(T) = \beta > 0$ but $r(T)$ does not belong to $\sigma(T)$.

3. IRREDUCIBLE POSITIVE OPERATORS

In this section we shall show that the spectral properties of a large class of irreducible positive operators can be derived from the theory of strict semi-simple locally compact semi-algebras (see section III.2).

Throughout this section we shall suppose that E is a real ordered Banach space and that the cone K of E contains non-zero elements. The ordering induced by K in E will be denoted by $\leq$. Thus we write $x \leq y$ (or equivalently $y \geq x$) if $y - x \in K$.

A cone K_o in E is said to be a *full subcone* of K if

(i) $K_o \subset K$,

(ii) from $0 \leq x \leq k$ for some $k \in K_o$ it follows that $x \in K_o$.

A positive linear operator T on E is said to be *reducible* if there exists a non-zero full subcone K_o of K invariant under T such that K_o is not a total subset of E. An *irreducible operator* is a non-zero positive operator that is not reducible. Thus a non-zero positive operator T is irreducible if every T-invariant non-zero full subcone of K is a total subset of E.

The concept of an irreducible operator is abstracted from the notion of an irreducible non-negative matrix ([12], Chapter XIII). To see this, consider the n-dimensional Euclidean space $\mathbb{R}^n$ with its usual ordering. The cone of $\mathbb{R}^n$ is the set

$$C = \{(x_1,\ldots,x_n): x_i \geq 0\}.$$

Let σ be a subset of $\{1,2,\ldots,n\}$, and let

$$C_\sigma = \{(x_1,\ldots,x_n) \in C: x_j = 0 \ (j \in \sigma)\}.$$

Then C_σ is a full subcone of C and it is not difficult to show that each full subcone of C is of this form. If we identify in the usual way an $n \times n$-matrix

$$T = (t_{ij})$$

with a linear operator on $\mathbb{R}^n$, then T is positive if and only if the matrix has non-negative entries, and in that case T is reducible if and only if there exists a partition of $\{1,2,\ldots,n\}$ into two disjoint non-empty sets σ and σ' such that

$$t_{ij} = 0 \quad (i \in \sigma,\ j \in \sigma'). \tag{1}$$

In fact for a non-negative matrix (t_{ij}), the condition (1) implies that the corresponding operator T leaves invariant the full subcone C_σ and conversely. In [12], Chapter XIII condition (1) is used to define the concept of a reducible non-negative matrix.

A linear operator T on an ordered Banach space E is called *strictly positive* if for each non-zero x in K

$$0 \neq Tx \in K.$$

Clearly, a strictly positive operator is positive and non-zero. Further, any bounded irreducible operator on E is strictly positive (see the next lemma), but the converse is not true.

3.1 LEMMA. *Let* T *and* S *be non-zero bounded positive operators on* E *such that the operator*

$$TS - ST$$

is positive, and let T *be irreducible. Then* S *is strictly positive. In particular,* T *is strictly positive, if* T *is irreducible.*

PROOF. Let $K_o = \{x \in K: Sx = 0\}$. In order to prove that S is strictly positive, we have to show that K_o consists of the zero element only. Clearly, K_o is a full subcone of K. Since TS - ST is a positive operator on E,

$$-STx \geq 0 \quad (x \in K_o).$$

On the other hand S and T are both positive operators hence, since $K_o \subset K$,

$$STx \geq 0 \quad (x \in K_o).$$

Both formulas together imply that STx = 0 for each x in K_o. In other words K_o is invariant under T. But T is irreducible. Thus $K_o = (0)$ or K_o is a total subset of E. Since S is zero on K_o, the latter implies that S is zero on E. Contradiction, thus K_o consists of the zero element only.

If we take S to be T, then all the conditions of the lemma are satisfied, hence T is strictly positive.

3.2 THEOREM. *Let T be a bounded irreducible positive linear operator on E, and suppose that the peripheral spectrum of T is a set of poles of T. Then A(T) is locally compact, semi-simple and strict. Further*

(i) $0 < r(T) \in \sigma(T)$,

(ii) $\mathrm{Per}\sigma(T)$ *is a set of simple poles of* T,

(iii) *there exists* $0 \neq u \in K$ *such that*

$$Tu = r(T)u.$$

PROOF. The cone K of E is a T-invariant non-zero full subcone of K. Hence, since T is irreducible, K is a total subset of E, and it follows that T satisfies the conditions of Theorem 2.1. Among other things this implies that (iii) holds and that

$$r(T) \in \sigma(T) \tag{2}$$

Further, we can repeat the arguments of the first part of the proof of Theorem 2.1, to show that A(T) is strict and that the elements of A(T) are positive operators.

From the conditions on the spectrum of T and formula (2) we see that T satisfies the hypotheses of Theorem II.1.1. Thus A(T) is locally compact.

To prove that A(T) is semi-simple, take $0 \neq S \in A(T)$. Then S is a non-zero bounded positive operator on E. Since S commutes with T, the operator S satisfies the conditions of Lemma 3.1, and hence S is strictly positive. Take $x \neq 0$ in K. Then, since S is strictly positive, $0 \neq Sx \in K$, and hence, again using the strict positivity of S, we obtain that $S^2x \neq 0$. This shows that $S^2 \neq 0$, and we can apply Proposition III.1.2 to prove that A(T) is semi-simple.

Finally we observe that (i) and (ii) now follow from Theorem III.2.1. This completes the proof of the theorem.

Consider the eigenspace

$$N = \{x \in E: Tx = r(T)x\}.$$

We shall prove that under certain extra conditions on T the space N is one-dimensional. The proof of this is based on two lemmas.

3.3 LEMMA. *Let* T *and* S *be bounded positive operators on* E *such that* T *is irreducible and* S *is strictly positive. Further suppose that the operator*

$$S - TS$$

is positive. Then S *is irreducible.*

PROOF. Let K_o be a non-zero S-invariant full subcone of K. Put

$$K_1 = \{x: 0 \leq x \leq Sk \text{ for some } k \in K_o\}.$$

Then K_1 is a full subcone of K and

$$SK_o \subset K_1 \subset K_o. \tag{3}$$

Since S - TS is a positive operator, we have

$$0 \leq TSk \leq Sk \quad (k \in K_o),$$

and hence it follows that K_1 is T-invariant. But T is irreducible. Thus either $K_1 = (0)$ or K_1 is a total subset of K. Since S is strictly positive and $K_o \neq (0)$, SK_o and hence K_1 contain non-zero elements (see formula (3)). Thus K_1 is a total subset of E, and therefore, again using formula (3), the same is true for K_o. This shows that S is irreducible.

3.4 LEMMA. *Suppose that the identity operator* I *on* E *is irreducible, and let*

$$K \cap \{x \in E: \|x\| \leq 1\}$$

be compact. Then E *is one dimensional.*

PROOF. We begin by proving that a non-zero element f of the dual cone (see section 2) K' of K is strictly positive, that is, we shall show that

$$f(x) > 0 \quad (0 \neq x \in K'). \tag{4}$$

Take a fixed non-zero element u in K. Consider the operator S on E defined by

$$Sx = f(x)u.$$

Since u and f are non-zero elements in K and K' respectively, S is a non-zero bounded positive operator on E. If we take T to be I, then T and S satisfy the conditions of Lemma 3.1. Hence S is strictly positive and thus (4) holds.

Take a fixed non-zero element f_o in K', and let

$$X = \{x \in K: f_o(x) = 1\}.$$

Formula (4) together with the compactness condition on K implies that X is a compact set. Now let f be an arbitrary element of K'. Define

$$\mu = \sup \{\lambda \in \mathbb{R}: f - \lambda f_o \in K'\}.$$

Then $0 \leq \mu < +\infty$ and $f - \mu f_o \in K'$. Suppose that $f \neq \mu f_o$. According to the result of the first paragraph this assumption implies that $f - \mu f_o$ is strictly positive. In particular, since $0 \notin X$,

$$(f - \mu f_o)(x) > 0 \quad (x \in X).$$

But X is compact. Hence there exists $\varepsilon > 0$ such that

$$(f - \mu f_o)(x) \geq \varepsilon \quad (x \in X),$$

and therefore

$$(f - \mu f_o)(x) \geq \varepsilon f_o(x) \quad (x \in K).$$

This implies that $f - (\mu + \varepsilon)f_o \in K'$, contradicting the definition of μ. Thus $f = \mu f_o$. But then we can use formula (5) of section 2 to show that

$$K = \{x \in E: f_o(x) \geq 0\}.$$

If $f_o(x) = 0$, then the preceding formula shows that $x \in K \cap (-K)$, and so $x = 0$. Hence the null space of f_o consists of the zero element only. Therefore E is one dimensional.

Let T be a bounded linear operator on an arbitrary Banach space. Recall that a pole λ of T is said to be of finite rank if the spectral projection associated with λ has finite dimensional range. The next theorem gives some more information about the spectral properties of a subclass of the class of operators considered in Theorem 3.2.

3.5 THEOREM. *Let* T *be a bounded irreducible positive linear operator on* E*, and suppose that the peripheral spectrum of* T *is a set of poles of* T *of finite rank. Then the eigenspace*

$$N = \{x \in E: Tx = r(T)x\}$$

is one dimensional.

PROOF. From Theorem 3.2 we know that A(T) is locally compact, semi-simple and strict. Further we have $0 < r(T) \in \sigma(T)$. By multiplying T with a suitable positive number, we may suppose without loss of generality that $r(T) = 1$.

Let Q be the spectral idempotent associated with the spectral set $\{1\}$. From the theory of semi-simple strict locally compact semi-algebras (see section III.2) we know that Q is the complexification of the idempotent

$$P = \lim_{n\to\infty} \frac{1}{n} (T + \dots + T^n). \tag{5}$$

Since 1 is a simple pole of T (see Theorem 3.2(ii)), it follows that

$$N = \{x \in E: Px = x\}.$$

Thus the dimension of N is at least one.

Observe that formula (5) implies that P is a positive operator such that $TP - PT$ is the zero operator. Hence, by Lemma 3.1, the operator P is strictly positive. But then, since $P = TP$ (cf., Proposition I.2.1), we can apply Lemma 3.3 to show that P is irreducible.

Let $C = PK$. Then C is a non-zero closed cone in N and therefore N is a real ordered Banach space. Since N is finite dimensional

$$C \cap \{x \in N: \|x\| \leq 1\}$$

is compact. Hence in order to prove that N is one dimensional, it suffices to show that the identity operator I_N on N is an irreducible positive operator on N (Lemma 3.4). Let C_o be a non-zero full subcone of C. Define

$$K_o = \{x \in K: Px \in C_o\}.$$

Then K_o is a full subcone of K. Further, since $C_o \subset K_o$, the cone K_o is non-zero and P-invariant. This implies that K_o is a total subset of E, and hence by the linearity and continuity of P, the cone C_o is a total subset of N. So any non-zero full subcone of C is a total subset of N. Therefore I_N is irreducible and the proof is complete.

NOTES ON CHAPTER IV

1. A cone K in a Banach space is said to be *normal* if there exists $\gamma \geq 1$ such that

$$\|x\| \leq \gamma\|x + y\| \quad (x,y \in K).$$

Thus one may rephrase Lemma 1.5 by saying that any strict locally compact semi-algebra is a normal cone. In Theorem 1.7 it is the normality of the semi-algebra that makes it possible to prove that the pole $r(T)$ is of maximal order in $\text{Per}\sigma(T)$. The arguments of this part of the proof of Theorem 1.7 are taken from section 2 in the Appendix of [31].

2. As shown in section 2, Theorem 1.7 yields a semi-algebraic proof of the Krein-Rutman theorem. Another proof of this theorem by using semi-algebra techniques is given by Bonsall in [3]. Bonsall's proof is based on the fact that a compact linear operator acts compactly on its centraliser.

3. It is interesting to observe that with an obvious modification Theorem 2.1(i), (ii) and (iii) also hold for the adjoint T' of T. Since the spectral properties of T' are the same as those of T, this is trivially true for (i) and (ii). To prove (iii) for T', recall that in the proof of Theorem 2.1 we established the existence of a non-zero positive operator S such that

$$TS = r(T)S.$$

By taking adjoints and using the fact that S commutes with T, we obtain

$$T'S' = r(T)S'. \tag{1}$$

Since S is non-zero, formula (5) of section 2 implies that there exists an element $g \in K'$ such that $f = S'g \neq 0$. But then f is a non-zero element in K' and, by formula (1),

$$T'f = r(T)f'.$$

4. In sections 2 and 3 all positive operators are required to be bounded. Therefore it is interesting to know that under certain extra conditions on the ordered Banach space E positivity of the operator implies boundedness. For instance this happens if

$$E = K - K.$$

In particular this is true if E is a Banach lattice (see [31], section V.5 for details).

5. Most generalizations of the Krein-Rutman theorem deal with the compactness condition on the positive operator T. It is well-known that T does not have to be compact. It is enough to suppose that T satisfies certain spectral conditions which may be considerably weaker than those required in Theorem 2.1. The best result in this direction replaces the compactness condition on T by the requirement that $\text{Per}\sigma(T)$ contains a

non-empty spectral set (see Lemma 2 (4,2) in [13]).

6. For a real bounded linear operator T with positive spectral radius and such that Perσ(T) is a set of poles of T of finite rank Theorem 2.4 is due to D.W. Sasser (see Theorems 2 and 3 in [30]). For a real compact linear operator T with $r(T) > 1$ Theorem 2.7(iii) is Theorem 1 in T.E.S. Raghavan [27]. In its full generality Theorem 2.4 is new and the same is true for the proof given here. Theorem 2.7 (in its full generality) has appeared in [18].

7. From the example mentioned at the end of section 2, it is easy to see that Theorem 2.7 does not remain true if the condition $r(T) \geq 1$ is replaced by the condition $r(T) = \beta$ with $0 < \beta < 1$. If $r(T) \geq 1$ is replaced by $r(T) = 0$, then (i) and (ii) of Theorem 2.7 hold trivially. To prove that (iii) also holds in that case, we observe that $r(T) = 0$ implies that T is nilpotent. Let k be the order of nilpotence. Then T^{k-1} is non-zero, hence, since C is a total subset of E, the set $T^{k-1}C$ contains non-zero elements. Take $u \neq 0$ in this set. Since C is invariant under T, the element $u \in C$. Further

$$Tu = 0 = r(T)u.$$

This shows that (iii) of Theorem 2.7 holds for $r(T) = 0$.

8. The result mentioned in Note 5 on Chapter II shows that for compact operators with positive spectral radius Theorem 2.1 cannot be improved without extra conditions on the cone K. A similar remark holds for Theorem 3.5 (see Satz 3.3 in L. Elsner [11]).

REFERENCES

1. H. Bart, M.A. Kaashoek, H.G.J. Pijls, W.J. de Schipper and J. de Vries, *Colloquium semi-algebras and positive operators* (in Dutch), Mathematical Centre Syllabus 11, Amsterdam (1971).

2. F.F. Bonsall, Locally compact semi-algebras, *Proc. London Math. Soc.* 13 (1963), 51-70.

3. F.F. Bonsall, Compact linear operators from an algebraic standpoint, *Glasgow Math. J.* 8 (1967), 41-49.

4. F.F. Bonsall and B.J. Tomiuk, The semi-algebra generated by a compact linear operator, *Proc. Edinburgh Math. Soc.* 14 (1965), 177-195.

5. G. Brown and W. Moran, Idempotents of compact monothetic semigroups, *Proc. London Math. Soc.* 22 (1971), 203-216.

6. G. Brown and W. Moran, An unusual compact monothetic semigroup, *Bulletin London Math. Soc.* 3 (1971), 291-296.

7. G. Brown and W. Moran, The idempotent semigroups of compact monothetic semigroups, *Proc. Royal Irish Acad., Section A,* 72 (1972), 17-33.

8. N. Dunford, Spectral theory. I. Convergence to projections, *Trans. Amer. Math. Soc.* 54 (1943), 185-217.

9. N. Dunford and J.T. Schwartz, *Linear Operators,* parts I and II, New York (1958, 1963).

10. R. Ellis, Locally compact transformation groups, *Duke Math. J.* 24 (1957), 119-126.

11. L. Elsner, Monotonie und Randspektrum bei vollstetigen Operatoren, *Arch. Rational Mech. Anal.* 36 (1970), 356-365.

12. F.R. Gantmacher, *Matrizenrechnung* II, Berlin (1966).

13. K. Georg, *Zur Spektraltheorie kegelinvarianter Operatoren,* Gesellschaft für Mathematik und Datenverarbeitung, Nr. 9, Bonn (1969).

14. T.A. Gillespie and T.T. West, Operators generating weakly compact groups, *Indiana Univ. Math. J.* 21 (1972), 671-688.

15. E. Hewitt and K.A. Ross, *Abstract harmonic analysis,* Vol. I, Berlin (1963).

16. E. Hille and R.S. Phillips, *Functional analysis and semigroups*, Amer. Math. Soc. Colloquium Publications, Vol. 31, Providence (1957).

17. M.A. Kaashoek, On the peripheral spectrum of an element in a strict closed semi-algebra, *Colloquia Mathematica Societatis János Bolyai* 5. Hilbert space operators, Tihany (Hungary), 1970, Amsterdam (1971), 319-332.

18. M.A. Kaashoek, A note on the spectral properties of linear operators leaving invariant a convex set, *Proc. Acad. Sci. Amsterdam* A 76 (1973), 254-262.

19. M.A. Kaashoek and T.T. West, Locally compact monothetic semi-algebras, *Proc. London Math. Soc.* 18 (1968), 428-438.

20. M.A. Kaashoek and T.T. West, Semi-simple locally compact monothetic semi-algebras, *Proc. Edinburgh Math. Soc.* 16 (1969), 215-219.

21. M.A. Kaashoek and T.T. West, Compact semigroups in commutative Banach algebras, *Proc. Camb. Phil. Soc.* 66 (1969), 265-274.

22. M.G. Krein and M.A. Rutman, Linear operators leaving invariant a cone in a Banach space, *Uspehi Mat. Nauk* 3 (1948), 3-95 (Russian); *Amer. Math. Soc. Transl.* 26 (1950).

23. K. de Leeuw and I. Glicksberg, Applications of almost periodic compactifications, *Acta math.* 105 (1961), 63-97.

24. K. Numakura, On bicompact semigroups, *Math. J. Okayama Univ.* 1 (1952), 99-108.

25. A.B. Paalman-de Miranda, *Topological semigroups*, Mathematical Centre Tracts 11, Amsterdam (1964).

26. R.R. Phelps, *Lectures on Choquet's theorem*, Van Nostrand Mathematical Studies 7, New York (1966).

27. T.E.S. Raghavan, On linear operators leaving a convex set invariant in normed linear spaces, *Mathematika* 17 (1970), 57-62.

28. C.E. Rickart, *General theory of Banach algebras*, New York (1960).

29. W. Rudin, *Fourier analysis on groups*, New York (1967).

30. D.W. Sasser, Quasi-positive operators, *Pacific J. Math.* 14 (1964), 1029-1037.

31. H.H. Schaefer, *Topological vector spaces,* New York (1966).

32. A.E. Taylor, *Introduction to functional analysis,* New York (1958).

33. T.T. West, Weakly compact monothetic semigroups of operators in Banach spaces, *Proc. Royal Irish Acad.,* Section A, 67 (1968), 27-37.

SUBJECT INDEX